THE
SCIENCE CLASS
YOU WISH YOU HAD

THE
SCIENCE CLASS
YOU WISH YOU HAD

The Seven Greatest
Scientific Discoveries in History
and the People Who Made Them

REVISED AND UPDATED

David Eliot Brody
and Arnold R. Brody, PhD

A PERIGEE BOOK

A PERIGEE BOOK
Published by the Penguin Group
Penguin Group (USA) Inc.
375 Hudson Street, New York, New York 10014, USA

USA | Canada | UK | Ireland | Australia | New Zealand | India | South Africa | China

Penguin Books Ltd., Registered Offices: 80 Strand, London WC2R 0RL, England
For more information about the Penguin Group, visit penguin.com.

Revised Perigee trade paperback ISBN: 978-0-399-16032-5

The Library of Congress has cataloged the first Perigee edition as follows:

Brody, David Eliot.
The science class you wish you had : the seven greatest scientific discoveries in history
and the people who made them / David Eliot Brody and Arnold R. Brody.
p.   cm.
"A Perigee Book."
Includes bibliographic references and index.
ISBN 978-0-399-52313-7
1. Discoveries in science—History.   I. Brody, Arnold R.   II. Title.
Q180.55.D57B76   1997
500—DC21   96-45097 CIP

PUBLISHING HISTORY
First Perigee trade paperback edition / August 1997
Revised Perigee trade paperback edition / August 2013

PRINTED IN THE UNITED STATES OF AMERICA

10   9   8   7   6   5   4   3   2   1

Text design by Tiffany Estreicher

While the authors have made every effort to provide accurate telephone numbers, Internet addresses,
and other contact information at the time of publication, neither the publisher nor the authors
assume any responsibility for errors, or for changes that occur after publication. Further,
the publisher does not have any control over and does not assume any responsibility
for author or third-party websites or their content.

Most Perigee books are available at special quantity discounts for bulk purchases for sales
promotions, premiums, fund-raising, or educational use. Special books, or book excerpts, can
also be created to fit specific needs. For details, write: Special.Markets@us.penguingroup.com.

ALWAYS LEARNING   PEARSON

CONTENTS

ACKNOWLEDGMENTS FOR THE ORIGINAL EDITION (1997) AND THE SECOND EDITION (2013)

This project would not have been possible without the assistance of a talented and generous group of people. First, our cousin, Stephen Karon, understood astronomy, physics and electromagnetic fields and forces before he was ten years old. We thank him for his careful review of the entire manuscript and his valuable contributions. Kenneth and Bette Brody, both chemistry teachers and experts in the history of science, took time to read several chapters critically and provided numerous suggestions to improve them. They also directed us to useful sources.

In the early stages of our research, Joseph Broz, PhD, a physicist with broad expertise and with much insight into the relationship between science and society, shared with us his knowledge of physics, relativity, and the history of science, and directed us to a number of excellent sources. Robert C. Amme, PhD, professor of physics at the University of Denver for the past thirty years, spent many hours reviewing the drafts of Part Three (relativity) and explaining its subject. David's colleagues Becke McGee and Brad McKim also read Part Three and gave us several constructive suggestions that were incorporated into the manuscript. In combination, these four people helped us present relativity in a manner that will, we hope, bring the seemingly untouchable subject of space and time down to earth for the reader.

We thank Peter Jordan for his comments on Part Five (evolution), for providing his expertise in geology and plate tectonics, and for pointing out the notable scientific discoveries and contributions made by his Scottish countrymen.

David's wife, Susan, read the entire manuscript and provided many creative comments. We thank her for those comments as well as for her patience, support, and encouragement. Arnold's wife, Toby, reviewed and listened intently to the additions prepared for the second edition and offered ideas and encouragement throughout the process. We thank the Penrose Library at the University of Denver and its staff members for use of the rich depository of science research materials. In addition, the University of Colorado (Norlin Library), the branches of the city of Denver library system, and the Denver Museum of Natural History were very cooperative in providing their extensive and comprehensive resources and information.

We owe much thanks to our agent, Elizabeth Kaplan of the Ellen Levine Literary Agency in New York, who made this entire book possible by her early and accurate advice on the approach and direction it should take and by finding the right publisher to bring it to fruition. Our editor at the Berkley Publishing Group for the original 1997 edition, Suzanne Bober, was highly supportive of this project from the beginning of our association with her, and expertly guided it through all the stages in the publishing process. We greatly appreciate her comprehensive and thoughtful editing of the manuscript.

We also thank our editor at Perigee Books/Penguin Group for the second edition, Jeanette Shaw, and our copyeditor, Candace Levy.

Bonnie Rothschild did an excellent job of typing and revising the many drafts of the manuscript that were necessary over the years during which it was in the making. We thank her for the quality of her work and her promptness and reliability in always meeting our deadlines, often resulting in her working all hours of the day and all days of the week. Finally, we thank Janice Gilmore for her many hours of work in 2011 and 2012 typing and editing the revised manuscript for the second edition.

—David E. Brody and Arnold R. Brody

In Memoriam . . . Martin Rodbell, PhD

"The fundamental thing is the opportunity to be creative," said Martin Rodbell, PhD, recipient of the Nobel Prize in physiology or medicine in 1994. Sadly, Dr. Rodbell passed away on December 7, 1998, at the age of seventy-three. He had discovered a central component of the communication system that allows cells to read external and internal signals from the millions of molecules in which our cells are immersed. The so-called G-proteins that Rodbell discovered transmit signals from hormones and proteins that control normal and abnormal functions, such as those for cell growth and neurotransmission. Aberrations of G-proteins are the basis of a variety of disease states, including cancer, heart defects, and cholera. Indeed, in 2012, the Nobel Prize in chemistry was awarded to Drs. Robert Lefkowitz of Duke University and Brian Kobilka of Stanford University. These brilliant investigators followed up on Rodbell's prize-winning work and determined at the molecular and atomic levels how the G-proteins bind inside cells, thus triggering the cascades of signals that control cell functions in health and disease.

Dr. Rodbell was born in Baltimore in 1925 and graduated from Baltimore City College. He spent a short time at Johns Hopkins University but was drafted and joined the U.S. Navy in 1943. He returned to Johns Hopkins after World War II and earned a bachelor of arts degree in biology in 1949. In 1954, he received a doctorate from the University of Washington and carried out postdoctoral training at the University of Illinois. In 1961, he moved to the National Institutes of Health, where he carried out his ground-breaking work on cell-signaling mechanisms. From 1985 to 1989, Rodbell was the scientific director of the National Institute of Environmental Health Sciences, where Arnold R. Brody was fortunate to be his colleague.

Rodbell had won numerous awards for his research in addition to the Nobel Prize, including the Jacobeus Award and the Gairdner International Award, and he was a member of the National Academy of Sciences. Rodbell's work is not presented in our book as one of the "Seven Greatest Scientific Discoveries in History." However, his work does sit at the top of what we have called the "Six Levels of Scientific Achievement" (see page 359), which is where modern scientists function in today's world. We are honored and grateful that Dr. Rodbell read the first edition of our book and wrote the foreword in which he added his perspectives on our work and his own research and placed them in the context of the seven greatest scientific discoveries.

ORIGINAL FOREWORD
BY DR. MARTIN RODBELL

In this book's view, science can be likened to a beautiful tapestry woven with different threads that have common features, but also demonstrate the extraordinary diversity of human thought in the creative process. It is an important book to read, whether the reader is completely ignorant of science or is a professional scientist. It combines the rare talents of two individuals, of disparate backgrounds: a scientist (Arnold Brody) and a practicing lawyer (David Brody). Their breadth of knowledge, keen sense of social responsibility, and their gift for writing beautiful prose combine to reveal not only the course of scientific discovery but the aspirations and the frailties of the participants in the context of the social and moral climate in which the important discoveries were made.

Seven pillars of scientific thought and accomplishment are discussed in this book—discoveries that encompass five hundred years—from the Renaissance to the mid-twentieth century. The authors describe these discoveries in a manner that makes science and the history of science clear and understandable. We learn that our planet is like a speck of dust in a vast desert; our sun, around which we revolve, is a minor star among billions of others. Life may not have begun on the planet Earth; the vital substances in meteorites spinning through the vastness of space may have deposited the seeds of life during a fruitful period of the planet's formation from clouds of hot gas. We are indeed an evolutionary product of chance, a late event in the development of multicellular organisms. The complexity of form and function is not unique to humans but has evolved to allow us to reason. Yet reason and the rigors of scientific thought came late in human development, only within a fraction of the time it took for the appearance of the first hominid.

This book was written shortly before the advent of the second millennium. The last major discovery discussed here is the structure of DNA followed closely by the decoding principles showing that this structure is the repository of all the information encoding living material. This momentous event led to the development of molecular biology and enabled scientists to uncover a wealth of hitherto buried information.

During the past 50 years, mainly as a result of the largesse of the U.S. government, there has been an extraordinary increase in biological research, bolstered in a major way by the participation of physicists and chemists. One of the legacies of the Holocaust and the disruptions caused by World War II was the movement to the United States of dozens of Europe's most important scientists in biology and physics. The National Institutes of Health in Bethesda, Maryland, supported their efforts, encouraging post-doctoral training of MDs and PhDs in an atmosphere conducive to creative thought and action. A remarkable group of scientists arose well-trained with a broad perspective on biology and medicine. They filled American universities and medical schools. Western Europe and Japan, devastated by World War II, benefited by having their young scientists trained at American medical schools and the National Institutes of Health. The result has been an enormous cross-fertilization of ideas around the globe, stimulating what has become the most productive period in the history of physics and biology. Scientific research is now truly a global effort in which intellectual barriers are nonexistent and cultural differences have been minimized. Indeed, the global society of scientists—of every cultural and ethnic persuasion—epitomizes the true spirit of the quest for knowledge.

Success has brought, paradoxically, dramatic changes in the scientific establishment and the process of scientific discovery. Scientists in all developed countries are witnessing a leveling off or even a downturn in the amounts of public funds devoted to science. As this book illustrates, discovery is no longer in the hands of a few. Group science, spurred in part because of decreased funding, is the rule. As a consequence, discovery has become incremental rather than dramatically episodic. However, it must be emphasized that the problems faced in science are enormous. In my field alone—biological communication—there have been dramatic changes in perception and knowledge, which are the result of discoveries made by relatively few scientists. As a result of these basic discoveries, the complex nature of the communicating system in cells and organs has been revealed, and has brought into focus the daunting task of integrating such knowledge. Certainly this task will require a concerted, multi-discipline effort if we hope to understand how communication systems operate at the cellular and organ levels. Individual scientists, no matter how brilliant, cannot possibly provide the answers. Group effort may solve, perhaps, the most mysterious aspects of life—how the brain functions, the basis of the "mind," and how the process of communication within cells and organs dovetails with genetic information to secure a more sound basis for Darwin's Natural Selection theory.

Discovery in such fields will be incremental as scientists face the limits of knowledge that can be acquired from investigating complex systems. This does not mean that the "end of science" is approaching, as some have suggested. Truth may be ephemeral, but Science, the Flower of Mankind, will continue as long as the human spirit exists on this planet. Brilliantly, this book takes the reader through the sowing of this flower's seed, the story of its growth, the lives of the extraordinary individuals who nurtured it, and the magnificent blossoming that is science in the 20th century.

—Martin Rodbell, PhD
Scientist emeritus, Nobel Laureate (1994)
in Medicine or Physiology

INTRODUCTION

Can the average person really understand science? Does the average person *want* to know about science? Does science matter to us? The answer to these questions is a resounding YES!

For some of us, however, the mere memory of physics, chemistry, and biology classes in high school and college makes our eyes glaze over. Some concepts, equations, and other scientific details seemed dull and abstract and virtually impossible for the average person to understand. To many of us, science seemed to have little immediate relevance to our lives. Yet as we matured and headed out into the world, we found ourselves face to face with sophisticated computers at work and frequent headlines about matters of science—mapping the human genome, newly discovered planets and their moons, the discovery of the possibility of past life on Mars, and in 2009 and 2011 the discovery of two close relatives to humans, to name just a few. Scientific knowledge has not only become acceptable but has become a useful, essential, and inescapable part of our lives.

For some of us, our fascination with science began sometime in the 1950s or 1960s, when the Soviet Union launched *Sputnik* or when Neil Armstrong set foot on the moon, striking evidence of humankind's ability to apply scientific knowledge to accomplish an extraordinary goal. For others, all it took to become interested in science and to spark their imagination was witnessing the unending series of new scientific achievements and inventions that occurred during the last decades of the twentieth century and into the twenty-first: Venus landing, fiber optics, reading the DNA code, black holes, space stations, microchips and computers, microsurgery, evolution of the space shuttle, heart and organ transplants, an artificial heart, superconductivity, the discovery of other solar systems, personalized medicine, and much more.

You don't have to be a theoretical physicist to be awed by space exploration or curious about whether there is life on Mars or how the universe began. You don't have to be a biochemist to have an interest in the fundamental processes of life. It's impossible *not* to be curious about such matters. Scientific knowledge and discoveries are much too interesting and profound to be left only to scientists.

The Seven Greatest Scientific Discoveries in History . . .

All those phenomenal technological achievements and inventions we've witnessed and experienced over recent decades have a common thread: *They became possible as a result of the basic scientific discoveries in physics, chemistry, and biology (and their subdisciplines) made over the last four centuries.* This book identifies the greatest of those discoveries—seven findings so profound that essentially everything humankind understands in science is based on them:

Gravity and the basic laws of physics

The structure of the atom

The principle of relativity

The Big Bang and the formation of the universe

Evolution and the principle of natural selection

The cell and genetics

The structure of the DNA molecule

These are the discoveries of *what is*, not what has been invented or what has been modified or developed by humankind, such as computers, space flight, insecticides, and penicillin, all spectacular achievements; this book is *not* an examination of microchips or computers. It *is* a look at how the universe came into existence, including life, how it all works, and what it's made of, from the smallest elementary particle to the most enormous and distant galaxy.

We do not ask or answer the question of *why* the universe or life exists. As the famous theoretical physicist Stephen Hawking stated, that would require knowing "the mind of God." And we do not ask what *should be* because that's also outside of science's domain. We can leave such questions to philosophers and clergy.

The selection of these particular discoveries was based on a combination of criteria, including opinions of scholars and experts in science and the history of science. These seven discoveries form the foundation that supports the enormous body of scientific knowledge that's been built. Without an understanding of physics, there would be no Venus landings or Hubble telescope. If we did not know the structure and function of the atom, nuclear power plants and the threat of nuclear war would not exist. If the principles of genetics had not been discovered, farm production and the world's food supply would be greatly diminished. Without a thorough understanding of the DNA molecule, cures for Parkinson's disease, sickle-cell anemia, and hemophilia would not be on the horizon. Each breakthrough presented here marked a new epoch in scientific knowledge and often brought with it monumental ethical and philosophical debate and controversy. These discoveries not only unlocked the intellectual and material riches of the modern world but have had an enormous impact on our daily lives.

The first four of these seven discoveries deal primarily with physics in its broadest sense and with astronomy. The last three deal with biology, including its related fields and subdisciplines (such as medicine and biochemistry). The matter and the concepts involved in these discoveries are impossible to see with the naked eye (as with the atom, the cell, and DNA) or are unthinkably abstract (as in the case of gravity and relativity) or they embody and reflect an enormous event (Big Bang) or a process of unimaginable duration (evolution). Because these discoveries can't readily be seen, are so abstract, or cover an enormous space or time span, they conflict with our everyday experiences—our own common sense. As a result, neither the scientific community nor the general public realized the full significance of these discoveries at the time they were made because each discovery acquired its greatness only after withstanding intense scrutiny.

In each instance, to reach a full understanding of how the physical universe and life on this earth were created and developed or evolved to their present state, we had to give up some perceptions and assumptions about science and life—what we thought was common sense—and to approach each discovery with an open mind. Today, the greatest obstacle to full understanding is so-called wisdom of the past. We're plunked down "here and now" in the twenty-first century with all the myths and half-truths that we have dragged with us from prior centuries, amid the complex social and cultural context that developed over the last few thousand years. That body of beliefs and that culture originated well before *any* great scientific discoveries—before anyone had the knowledge to answer profound questions about the creation of the universe and life, before we discovered the Big Bang or the principle of evolution, and before we recognized the clear distinction between concepts based on creativity or our fertile imaginations and concepts based on reality.

One has to approach the study of science with an open mind. You cannot fully understand the concept of gravity until you realize it is more than a matter of semantics to distinguish between an object *falling* and being *pulled* to the ground. The same force that keeps our feet on the ground causes the Earth to revolve around the sun. Accepting the underlying principles of science will change your fundamental views of life, the universe, and the human race. For example, understanding evolution presents an entirely new perspective on the battles and conflicts throughout history that were the result of physical and cultural differences between people, a perspective that tends to knock down barriers.

. . . and the People Who Made Them

This book also describes the triumphs and tragedies, lives and motives of the extraordinary people who were central to those great discoveries. We discuss and recognize the work of hundreds of people, but give the primary credit to ten of them.

Part One	Gravity and physics	Isaac Newton
Part Two	Atom	Ernest Rutherford, Niels Bohr
Part Three	Relativity	Albert Einstein
Part Four	Big Bang	Edwin Hubble
Part Five	Evolution	Charles Darwin
Part Six	Cells and genetics	Walther Flemming, Gregor Mendel
Part Seven	DNA	Francis Crick, James Watson

Their attitudes, personalities, and sheer determination, in combination with their ideas, led to these discoveries. For example, in 1543 Nicolaus Copernicus reluctantly published his book stating that the sun, not the Earth, is the center of the universe. This idea touched off a revolution in thought that carried with great force and controversy into the next century. Galileo, the Italian mathematician, astronomer, and physicist, spent years attempting to prove the truth of Copernicus's theory. When Galileo put that proof in print in 1632 in *Dialogue Concerning the Two Chief World Systems*, the church determined he'd violated its prohibition against teaching Copernican views, and pronounced the book to be heresy. Ultimately, Galileo was forced to retract his words publicly and vow never to teach Copernican theory again or risk being tortured and burned alive at the stake.

Isaac Newton, the father of physics, led the world into the next scientific era. He was born in 1642, the year Galileo died. Newton's father had also died earlier that year, and the infant was left by his mother in his grandmother's care. Years later, with the university closed during the great plague of London, Newton retreated to his secret world at Woolsthorpe, England, where he invented calculus and conceived the idea of universal gravitation. Yet he didn't publish his work for more than twenty years. Newton's work was finally revealed in 1687 in his book, the *Principia*, still considered to be the greatest single scientific intellectual feat in history. As with any fundamental change in thought, this new idea called gravity was accepted only slowly. "You have not explained why gravity acts," the critics accused. So, like Copernicus, Galileo, and others before and after, Newton had to fight another battle in the war for rational thought.

In the mid-1800s a monumental battle was emerging in the field of biology, in which Charles Darwin became the commanding general. His proposition was unique and dealt with an issue that had never before been the subject of the scientific method of investigation: the concept of life itself, its origin and evolution—biology and organic matter, not the remote lifeless orbs studied by Copernicus, Galileo, and Newton. Because of the emotional elements of Darwin's subject, this English gentleman found himself the focal point of the greatest controversy of the nineteenth century, one that continues unabated into modern times, from the Scopes monkey trial in 1925 to today's heated confrontations involving creationism versus evolution.

In many cases the social and political forces that these scientists faced influenced and often inhibited recognition of the importance of their discoveries. Science didn't make these people famous. Rather, these are the people who made science. This is history as well as science, for the two are inextricably linked, bound together with the economic, political, military, and religious strands that tie civilization into a coherent picture.

Yes, science can be fascinating. These seven discoveries have now, in our lifetime, culminated in the most incredible and pervasive scientific and technological revolution that could be imagined. Whether we approve of it or not, we're swept up in that revolution and the resulting culture—unless you live in a cave. Even then, you could have HBO and a smartphone. So, yes, not only is science fascinating but it matters to us because it is our lives. "The present epoch is a major crossroads for our civilization and perhaps for our species," said Carl Sagan (1934–1996) in *Cosmos*. "Whatever road we take, our fate is indissolubly bound up with science. It is essential as a matter of simple survival for us to understand science." Knowing science means knowing life. It means feeling more comfortable with our everyday lives, and using science and technology to accomplish goals. Science is part of our culture and heritage. It is for the masses, not merely ivory tower intellectuals.

We are extraordinarily fortunate to be living in this exciting transitional period, unique in all of history—one in which we're going from ignorance to knowledge, from questions to answers, from wonder to understanding. As we venture forward in the second millennium, a fog has lifted to show the people of this generation what those before us never knew. This is the objective of exploring the seven greatest scientific discoveries in history and the people who made them. We hope it will be *The Science Class You Wish You Had*, making science interesting and within every reader's grasp; rendering science intelligible for the nonexpert; and challenging personal myths, perceptions, and assumptions.

Back to a thought posed earlier: *The average person really can understand the great scientific discoveries.* In this entire book there are only a couple of equations—like Einstein's $E = mc^2$—and you won't be asked to use them. This book is about concepts and ideas, not algebra and calculus. Einstein himself said, "The whole of science is nothing more than a refinement of everyday thinking." Scientific discoveries are built on a foundation or framework supplied by mathematics, but words are all that are needed to understand them. By understanding the seven discoveries explored in the following pages and the role played by the scientists who made them, you will have accomplished a feat that would have astounded both you and your high school physics, chemistry, and biology teachers. This is the science class they should have been teaching. Finally, after many years, this book is what you needed. This is the science class you wish you had.

PART ONE

Gravity and the Basic Laws of Physics

In Focus

Ancient philosophies died slowly and stubbornly during the Renaissance. As Copernicus, Tycho, and Kepler gradually proved that Aristotle's and Ptolemy's theories about the universe were wrong, people realized the Earth was not its center and that it moves about the sun. But the forces of reason came up against the forces of politics and power; challenging the Roman Catholic Church's strict Aristotelian teachings cost Giordano Bruno his life and Galileo his freedom.

Then, in the latter part of the seventeenth century, in what is regarded as the greatest single intellectual accomplishment in scientific history, Isaac Newton discovered and described the law of universal gravitation and the basic laws of classical physics. Every particle of matter gravitationally attracts every other particle of matter with a force that is quantified in Newton's mathematical formula. As massive bodies that are made up of such particles, the sun attracts every planet in our solar system, and the Earth attracts the moon. If you were to let go of this book, the Earth would pull on it with the same invisible force.

Universal gravitation, the laws of motion, and the other quantitative rules developed by Newton marked the beginning of modern physics and formed the paradigm on which much of modern science has been built. Reason triumphed, and the world was changed forever.

Chapter One
THE REVOLUTIONS

Any revolution . . . begins with some definite act, often meant to purify corrupt practices and restore what some conservative radical imagines as a pristine state of things.
> —Charles C. Gillispie, *The Edge of Objectivity* (1960)

If you are a long-distance traveler . . . you can get your bearings by landmarks—mountains and rivers, buildings. . . . But . . . the emptiness and homogeneity of the sea . . . drove sailors to seek their bearings in the heavens, in the sun and moon and stars. . . . They sought skymarks to serve for seamarks. It is no wonder that astronomy became the handmaiden of the sailor, that the Age of Columbus ushered in the Age of Copernicus.
> —Daniel J. Boorstin, *The Discoverers* (1983)

The Black Death Kills Millions and Threatens a Return to the Dark Ages

Perfect Circles and Crystal Spheres

In the summer of 1347, the most devastating plague ever suffered by human-kind struck Europe—the bubonic plague caused by bacteria passed on to humans by fleas from rats and the pneumonic plague from the same bacteria passed person to person. It became known as the Black Death, so called because the victims' putrefying flesh appeared to blacken as blood coagu-lated in the final hours before death. The plague originated in China and was transmitted to Europeans through a grotesque early version of germ warfare when a Kipchak army (a tribe of the Mongol empire) attacked a Genoese trading post on the coast of the Black Sea and used its giant catapults to lob plague-infested corpses over the walls of the town. In his book *The Black Death*, historian Philip Ziegler described the next deadly steps:

> As fast as the rotting bodies arrived in their midst, the Genoese carried them through the town and dropped them in the sea. But few places are so vulnerable to disease as a besieged city. . . . Such inhabitants . . . realized that, even if they survived the plague, they would be far too few to resist a fresh Tartar onslaught. They took to their galleys and fled from the Black Sea towards the Mediterranean. With them traveled the plague.

Over the next four years, the disease spread in its first wave throughout Mediterranean ports in Sicily, North Africa, Italy, Spain, France, England, and finally all of Europe. The first symptoms were nosebleed, shivering, vomiting, headache, giddiness, intolerance to light, pain in the back and limbs, and eventually delirium. Quoting contemporary records, Ziegler described the physical appearance and characteristics of the plague as boils (buboes of the bubonic plague) "in the groin or armpits, some of which grew as large as a common apple" from where they "soon began to propagate and spread . . . in all directions indifferently . . . a token of approaching death." When the lungs became infected, the victim died within two days. Other cases dragged out for months. Wrote Ziegler,

> Everything about it was disgusting, so that the sick became objects more of detestation than of pity; . . . all the matter which exuded from their bodies let off an unbearable stench; sweat, excrement, spittle, breath, so foetid as to be overpowering; urine turbid, thick, black or red.

Efforts to check the disease included bonfires that were believed to disinfect the air, demonstrations of penitence, and persecutions of religious minorities. Twenty-five million people suffered those symptoms and perished—one-fourth of the population of Europe at the time, with some areas losing as much as 60 percent of their population. And this wasn't the end. There were five recurrences in the fourteenth century, and it would take 200 years before the European population would climb back to 100 million.

Nine hundred years earlier, after the fall of the Roman empire in the fifth century CE, the Church of Rome had become the center of Western Christendom. By the year 1300 the church had centralized its authority throughout Europe, becoming involved in secular elections, international conflicts, and all aspects of civic life. In Italy, the church had successfully taken up the task of restoring order in city-states that had fallen apart politically and economically. This success enhanced papal authority, and by the time of the Black Death, the Roman Catholic Church had consolidated its power. Turning to religion for comfort and salvation, laymen became more pious, and morality more rigid. There were ugly political and physical attacks on European Muslims and Jews, as they were given the choice of Christianity or death. Most chose to convert.

The devastation of the Black Death brought with it the end of the prosperity and optimism that had begun to emerge in Europe in the preceding decades. Suddenly, there was a climate of fear and anxiety and a continuous series of social and political upheavals throughout the continent. Because much of the workforce had fallen victim to the plague, the fields became neglected. Serfs escaped from their masters' control, and labor costs increased, leading to government efforts to control such costs, which culminated in the Peasants' Revolt of 1381. Goods became scarce as the economy, largely dependent on agriculture for survival, fell into a medieval version of a depression. Economic and political chaos resulted, as Europe threatened to sink back into the Dark Ages that had enveloped Western civilization during the years 450–750 CE.

At the time of the Black Death, the physical universe was defined by the principles laid down by the Greek philosopher Aristotle (384–322 BCE) and the Egyptian astronomer Claudius Ptolemy (100–170 CE). Over the centuries, those principles had been tempered by the medieval interpretation adopted by the church: *God* had set the heavens in perfect and eternal circular movement; our world, made of the four elements (earth, air, fire, water), was at the center. Eight concentric crystal spheres made of an immutable substance, and the other heavenly bodies that were carried or supported by the spheres, made up the heavens. One sphere carried the sun, one carried the moon, five separates spheres each carried one of the planets (other than the Earth) known at the time (Mars, Mercury, Jupiter, Venus, and Saturn), and the eighth sphere carried all the stars. Material on Earth decayed and died, while the rest of the universe was perfect and unchanging.

Aristotle's and Ptolemy's ideas about the universe ultimately had an impact that extended well beyond a mere physical description of the natural world. An elaborate and entrenched set of cultural beliefs developed, based directly on this vision of reality. Astrology was central to this body of thought. For centuries, astrologers—and the kings, queens, and emperors who relied on the astrologers' advice—looked upward to determine the location, "influence," and "character" of these heavenly bodies for everyday guidance. In turn, the lives of millions of people ruled by the leaders who reigned throughout much of the medieval period were affected by Aristotle's and Ptolemy's views of the universe.

Even the arbitrary compartmentalization of time into seven-day groupings is the direct result of this ancient concept of the heavenly bodies. Beginning during the Roman empire of 2,000 years ago, each day was dedicated to one of the seven "planets," though these included the sun and the moon, and excluded the Earth. The specific order of the days of the week was based on the "influence" of each planet on worldly affairs, not their distance from Earth. Thus institutionalized by astrology, our days of the week are named after the "planets" dating back to ancient Rome. As shown in the chart below, this connection remains obvious in English, and even more so in other languages.

Heavenly Body	English	French	Italian	Spanish
Sun	Sunday	dimanche	domenica	domingo
Moon	Monday	lundi	lunedì	lunes
Mars	Tuesday	mardi	martedì	martes
Mercury	Wednesday	mercredi	mercoledì	miércoles
Jupiter	Thursday	jeudi	giovedì	jueves
Venus	Friday	vendredi	venerdì	viernes
Saturn	Saturday	samedi	sabato	sábado

This artificial cluster of seven days, not tied to any regular movement of the heavenly bodies, was a result of imagination. Though it was an attempt to find regularity and included the idea that humans are influenced by invisible forces that act over great distances, the entire approach to figuring out the universe lacked any basis in fact.

Superstition and astrology ruled. Magic, witchcraft, and alchemy were popular. There was no science.

Commercial Needs Spark the Scientific Revolution

Johannes Müller and the Epicycles

By the early fifteenth century, Europe began showing signs of an impending transformation. Indeed, the legacy of the plague was an incentive to climb out of that dismal era as quickly as possible and bury it in the past. It set the stage for a bright new chapter—the Renaissance, the rebirth of Europe, characterized by the rediscovery of classical literature, the rebirth of art, and great interest in all intellectual endeavors. The Renaissance included the Scientific Revolution, the age of worldwide geographical exploration, and unprecedented efforts to develop technology aimed at facilitating economic and commercial enterprises.

Italy was first to fully recover from the Black Death. Because their country was ideally located for commercial trade between Europe and the Near East, Italians had been heavily engaged in world trade since the early 1300s, importing spices, perfume, silk, and other commodities from the Orient and brokering these items to northern European countries. As a result, Italy had developed sophisticated managerial systems, banking practices, and financial expertise that paved the way for the nation's renewed commercial preeminence in the 1400s. Florence was regarded as the world leader in international banking and trade. Italy's remarkable economic recovery following the medieval period also led to that nation's supremacy in scientific inquiry. Mathematics was used in construction, navigation, cartography, and surveying. Beginning in the 1400s, geometry, trigonometry, and algebra were refined and applied to a greater extent than ever previously imagined. Sixteen centuries earlier a few Greek philosophers had suggested the Earth was round—a view that had been buried deep in history all through the Middle Ages. But thanks to Paolo Toscanelli (1397–1482), the Italian cosmographer who provided Columbus with the map that guided him on his first voyage, and the new generation of explorers, which included Columbus, Vespucci, and Magellan, people began to realize the Earth was not a disk primarily composed of Europe and Asia. It was a sphere, largely covered with water, and all of its distant lands might be habitable.

With the use of mathematics in everyday life and the recognition that the Earth is round, ideas about the universe, including the movement of the moon and Earth and other planets, began to change. If a starting point for that change could be identified, it would be 1463, not in Italy, but in Germany, where an astronomer named Johannes Müller (1436–1476) wrote a book called the *Epitome* (published posthumously in 1496) in which he noted weaknesses in Ptolemy's Earth-centered or geocentric theories expressed in *The Almagest*. Müller was particularly critical of Ptolemy's convoluted "epicycles" or small loops within each planet's orbit that were needed to account for the movements of those planets in a universe in which the Earth stood still at the center. Although *The Almagest* had been written over 1,300 years earlier, it remained the authority on how the universe functioned—one of the great symbols of the unwavering reverence for the ideas espoused by Aristotle and Ptolemy. Müller's work was the first step toward renouncing the belief that the Earth was unmoving, immovable, and at the center of the universe, setting the stage for a revolution in astronomy.

Copernicus Topples Geocentricity and Dethrones Humankind

But Where Is the Wind?

We now know that it takes the Earth 365 days, five hours, forty-eight minutes, and forty-six seconds to make one trip around the sun. The ancient Egyptians calculated a year as 365 and *one-fourth days*—but that's eleven minutes, fourteen seconds too long. Thus when Julius Caesar adopted the Egyptian calendar in the 1st century CE, it led to a gradual disparity between the dates on the calendar and the seasons, so that by the mid-fifteenth century, the Julian calendar was 10 days off. Farmers and navigators noticed the problem. Also, the church had based Easter on the vernal equinox, which had originally been fixed as March 21, but was now occurring on March 11 because of the accumulating inaccuracy. In 1475, Pope Sixtus IV had asked Johannes Müller to undertake a study to determine the exact cause of the disparity. But Müller failed in that effort because he lacked the data, and it was virtually impossible to accumulate such data under the prevailing astronomical system of Aristotle and Ptolemy, with the moon and sun circling the immovable Earth and the eight crystalline spheres eternally rolling around the Earth. The continued adherence to Aristotle and Ptolemy was the bedrock of social stability, yet it was an impediment to an accurate Easter and the very birth of science.

In the meantime, Nicolaus Copernicus (1473–1543), was growing up in Torun, Poland, the youngest of four children of a prosperous merchant. Copernicus had a broad range of interests and talents and studied at several schools in Italy—astronomy at Bologna, medicine at Padua, and law at Farrara. He later lectured in various universities throughout Europe on mathematics, astronomy, and medicine, but spent most of his life as canon of Frauenburg, Poland, with his education and career under the auspices of the Catholic Church, the same as most scholars of the period.

In 1506 he began to develop an astronomical system based on his own observations and calculations on the movement of heavenly bodies. Like Müller, Copernicus soon recognized that Ptolemy's theories of a geocentric universe were not consistent with these observations. On the other hand, Copernicus knew that Ptolemy's writings were consistent with the Bible and that it does not appear Earth is rotating on its axis or revolving around the sun. The concept of gravity was not known yet, and Copernicus had no convincing argument to counter the universally accepted Aristotelian explanation that the reason objects fall to Earth is because matter is "naturally attracted" inward to the center of the universe, that is, the Earth. Copernicus didn't attempt to explain why objects did not "fall" off the Earth and toward the sun if the sun is in fact the center of the universe. Moreover, few were prepared to consider a theory that displaced Earth and humankind from the hub of the universe. The unassailable truth was that God put us in the center and on the largest object because vanity, fear, and the Bible demanded this correlation of location and size to importance.

When recruited by the pope's secretary in 1514 to investigate the same perplexing problem of the inaccurate calendar that had eluded Müller, Copernicus was faced with the dilemma of trying to explain his observations without refuting the popular and sacred notions of the day. Dissension between Catholics and Protestants made the church more vigilant about maintaining its teachings, thus increasing the chance Copernicus would be charged with impugning the church's authority. As a result, he declined the church's invitation and told them he could not explain the reason for the discrepancy until the relationship among the Earth, the sun, and the moon was more fully understood. He continued secretly to develop his theory while holding various governmental and ecclesiastic posts; from 1519 to 1521 he aided the reconstruction of Ermland, Poland, served as commissary for the Ermland diocese, and provided medical help to the community, and in 1522 he presented a plan to reform the local currency.

Convinced of the accuracy of his observations and encouraged by friends, Copernicus finally went public in 1530 when he presented a short outline of his sun-centered theory in a published article titled "The Commentariolus," prompting a mixed reaction. Over the next several years, he lectured on his findings, while his colleagues strongly encouraged him to publish the full statement of heliocentricity. Despite this support and despite Pope Clement VII's informal approval of the theory, Copernicus hesitated to publish it in its entirety because he remained reluctant to fully challenge the accepted view of the universe. Finally, in 1540, after much persuasion by his friends, Copernicus gave permission for the publication of his full and monumental work *De revolutionibus orbium coelestium, On the Revolutions of the Heavenly Orbs*, commonly called *The Revolutions*. Bedridden with apoplexy and paralysis, Copernicus was handed an advance copy of the book on May 24, 1543, and died later that day.

Divided into six long parts, *The Revolutions* gives a detailed explanation and justification for heliocentricity. It first established that the Earth is spherical, which many still doubted, and discussed the immensity of the heavens. Copernicus then described a system in which the six planets known at that time are centered around the sun instead of the Earth, with the Earth rotating and revolving about this "visible God." In Copernicus's system, planets' orbits remained in perfect circles on the crystal spheres, not on their actual elliptical orbits, because measurements were crude, and true speeds, sizes, and distances in space were totally unknown. Copernicus gave the first logical explanation why each planet (derived from the Greek word for "wanderer") orbits the sun in one revolution, and he did so without Ptolemy's numerous and unnecessary epicycles, which didn't even fit the concept of crystal spheres. The Copernican model of the universe also explained why the planets, viewed from Earth, alter their positions while the stars appear to remain fixed in their locations.

FIGURE 1-1

Solar System from *The Revolutions*

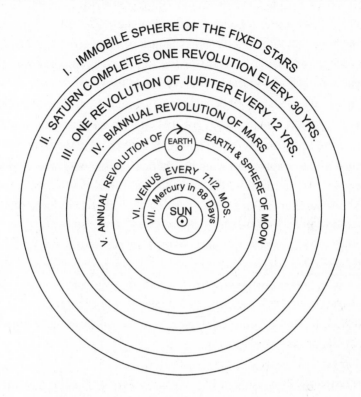

The Revolutions helped solve the calendar problem and showed why Ptolemy's theories seemed to be wrong. Yet few people, including astronomers, adopted Copernicus's theory. After all, he failed to explain why there was no great wind from this motion, why a shift of the stars' positions could not be detected, and why all objects (including people) were not hurled off the Earth's surface. Also, his theory conflicted with the Bible. Although the church did not officially denounce Copernicus when *The Revolutions* was published, there was a negative reaction from many religious leaders. The German theologian Philipp Melanchthon (1497–1560), a strong believer in astrology and demonology, had tried to prevent publication. The Protestant reformer John Calvin (1509–1564) pointed out that the Bible says the world cannot be moved, and the German preacher and biblical scholar Martin Luther (1483–1546) condemned Copernicus, stating, "The fool will turn the whole science of astronomy upside down." Even the wisest of men could not yet justify dethroning humankind and Earth from the center of the universe to a mere planet of no special significance. Thus, despite Copernicus and *The Revolutions*, Earth remained at the center of the universe for the time being, and it would remain the task of other independent-minded astronomers and mathematicians of the Renaissance to adopt his theory, refine it, and expand it to the level of irrefutability.

Copernicus's proposal of heliocentricity coupled with specific observations of the planets' movements was a bold effort to separate astronomy from philosophy and make it a true science. He rejected Aristotelian/Ptolemaic teachings about why bodies fall to their "natural place," he dethroned humankind and the Earth, and his work eventually led to the revision of people's concept of the universe—a fundamental change in thought known as the Copernican Revolution. This profound view also resulted in a new connotation for the term *revolution*—until Copernicus it meant only the physical movement of heavenly bodies. Perhaps most important, for the first time in history, he introduced the concept that the solar system can be viewed and studied as a structure independent from the stars.

In 1953, in his speech commemorating the 410th anniversary of the death of Copernicus, Albert Einstein said:

> Copernicus not only paved the way to modern astronomy; [he] also helped to bring about a decisive change in man's attitude toward the cosmos. Once it was recognized that the Earth was not the center of the world, but only one of the smaller planets, the illusions of the central significance of man himself became untenable. Hence, Copernicus, through his work and the greatness of his personality, taught man to be modest.

Indeed, as Martin Luther predicted, Copernicus did turn astronomy upside down. However, he was anything but a fool.

Tycho, the First Great Astronomical Observer, Maps the Universe

A Need for Compasses and Clocks

The Revolutions lay there on the periphery of astronomical thought, like an orphan, benign and unwanted. It was criticized by ecclesiastical leaders but not officially censored by the church, for the Roman Catholics had not yet begun their fight against heretics and Protestants in full force. A series of events then followed that were to immortalize Copernicus and give helio-centricity a greater significance than even its author had intended. The first was a partial eclipse of the sun in the year 1559, sixteen years after Coperni-cus's death. It had been predicted by the astronomical tables and had no particular significance in and of itself, except that it was witnessed by Tycho Brahe (1546–1601), then a thirteen-year-old student at the University of Copenhagen in his native Denmark, and made such an impression on him that it shifted the entire course of his life. He had already begun his studies in rhetoric and philosophy in preparation to follow the tradition of his aris-tocratic family and become a statesman but then became fascinated by man's ability to predict events such as this eclipse. He was soon preoccupied with observing the planets. However, his family and the church authorities would not tolerate this unconventional pastime. They forced Tycho to continue his general education and later his law studies at various universities, while he secretly devoted most of his time to his passion of observing the movement of the planets.

His desire to measure accurately the movement of heavenly bodies coincided with Denmark's need for precision in ocean navigation and an ever-increasing demand for accurate compasses and clocks throughout Europe. With this growing acceptance of math, technology, and astronomy, Tycho was finally permitted to devote all his time to observations without the earlier disapproval by his family and peers. He completed his long education in 1572 at age twenty-six. Then, on the evening of November 11, 1572, Tycho experienced another astronomical event that would help shape the history of science. He noticed a bright object near the constellation Cassiopeia that had never been there before. It was so unusually bright (brighter than Venus) that it was visible in daylight. In December, it began to fade but remained visible through March of 1573. Though there had been stories in history about the appearance of new shining bodies, people had continued to sub-scribe to the basic Aristotelian doctrine that although the region between the Earth and moon could change and decay, the heavens were immutable. In other words, new stars could not be created. In the past, such bright new objects were believed to be comets, since comets were thought to be located only in that area between the Earth and moon, that is, *within* the sphere of change and decay, not in violation of Aristotle's teachings. Thus the critical test of this new heavenly object of 1572 was whether movement could be detected. If it was moving, it was just another comet. If not, something was rotten in Denmark and in the rest of the universe, as conceived to date.

Over the several-month period that the object was visible, astronomers throughout Europe applied their crude measuring devices and techniques and concluded that this new light stood still. Likewise, Tycho observed and measured the phenomenon with his newest sextant, the most sophisticated in existence, and came to the same conclusion—it must be a star, not a comet. Aristotle's principles on the immutability of the heavens were wrong. Nevertheless, the alchemists and astrologers were quick to furnish their own explanations for the nova: It was thought to be a comet condensed from the rising vapors of human sin; made up of poisonous dust drifting down on people's heads; the cause of evil, bad weather, and pestilence.

Thus the long and awkward journey to the Renaissance continued, as Tycho Brahe complained of the "blind watchers of the sky" and wrote a book describing his detailed observations. Tycho didn't venture to guess *how* the star came into existence—it would take centuries before scientists would determine the explosive process causing novas. Instead, he concentrated on making accurate observations, and did it so well that in 1576 King Frederick II of Denmark granted him title to the 2,000-acre, three-mile-long island of Ven, in the sound off Copenhagen, for building and running an observatory at Denmark's expense. Tycho turned Ven into a monument reflecting his grandiose style. Completely funded by the state, he built a large home, an observatory, a chemical laboratory, clocks, sundials, globes, a mill for grinding corn, sixty fish ponds, flower gardens, an arboretum, a paper mill, a printing shop to produce and bind his manuscripts, a windmill and pump for water, and observational inventions and devices that were to become wonders of the world. This extravagant complex, which he chris-tened Uraniborg (meaning "heavenly castle"), was furnished with a staff of artisans to build the instruments, professional astronomers to record the universe, servants, and all the other workers required to run his little com-munity. Tycho spent the next twenty years at the observatory distinguishing himself as the pioneer of methodical, scientific observation. *All with the naked eye because the telescope had not yet been invented!*

Tycho Brahe never fully embraced Copernicus. Indeed, his own concept of the heavens included a modified version of geocentricity, including epicycles similar to Ptolemy's. He was not a creative genius, though he was the first person to apply higher math to observational astronomy when he made the significant discovery that a comet that he saw in 1577 orbited beyond the moon. His place in history is secured by a much more accurate mapping of the stars and planets and their movements than ever previously done, resulting in the revision of virtually every astronomical table that existed up to that time. He realized that astronomy needs precise data—for the calendar, navigation, or just for the sake of knowledge. Even Copernicus had failed to recognize the importance of this principle, having relied on the data of Ptolemy, the Greek astronomer Hipparchus, and others, which were largely inaccurate. Copernicus's *The Revolutions* records only twenty-seven observations. In contrast, Tycho undertook tens of thousands. Until Tycho, the emphasis was on qualitative and philosophical theory, not exact measurement. Tycho's work had immediate application for farmers, naviga-tors, and clockmakers, and was viewed as revealing God's work and God's plan.

But for Tycho to become an important link in the astronomy/physics chain leading from Copernicus to Isaac Newton, a new and unexpected application of Tycho's wealth of accurate information was first required. It would take a greater mind to somehow seize Tycho's data and discover the greater secrets hidden in his volumes. It would take another step to complete the story that Tycho Brahe began to write unwittingly with his fantastic devices that measured the movement of Earth, the other planets, and the stars.

Kepler Searches for God's Grand Scheme of the Universe and Discovers the Laws of Planetary Motion

A Mangy Dog . . . Scabs of Chronic Putrid Wounds and Musical Scales

"Let all keep silent and hark to Tycho," wrote the brilliant and eccentric Johannes Kepler (1571–1630). "For Tycho alone do I wait; he shall explain to me the order and arrangement of the orbits . . . Tycho possesses the best observations. . . . He only lacks the architect who would put all this to use." Copernicus had turned his back on Aristotle and Ptolemy and began to open the door to the future. Through that portal, the light of science and knowledge was emerging. Tycho proved that Copernicus was justified in standing firm at the threshold, and now Kepler was about to take the first step across it. He was born in Württemberg, Germany, the year before Tycho eyed the inspirational nova, and grew up in a poor family, the son of an irresponsible father, but distinguished himself as a good student and was given an education by the state. He left a great gift to humanity, particularly to historians and scientists who came after him, in the form of his prolific writings, most of which still exist. His notes, journals, and books consist of outpourings of all aspects of his life, including detailed chronicles of his childhood and health:

I was born premature, at thirty-two weeks . . .

1575 . . . I almost died of small pox, was in very ill health, and my hands were badly crippled

1577 . . . On my birthday I lost a tooth, breaking it off with a string which I pulled with my hands

1585–86 . . . During these two years, I suffered continually from skin ailments, often severe sores, often from the scabs of chronic putrid wounds in my feet which healed badly and kept breaking out again. On the middle finger of my right hand I had a worm, on the left, a huge sore

1587 . . . On April 4, I was attacked by a fever

1589 . . . I began to suffer terribly from headaches and a disturbance of my limbs. The mange took hold of me . . . then there was a dry disease

1591 . . . The cold brought on prolonged mange . . .

"That man has . . . a dog-like nature," said Kepler, writing about himself in third person.

His appearance is that of a little lap dog. His body is agile, wiry, and well-proportioned . . . so greedy that whatever his eyes chanced he grabbed. . . . He continually sought the good will of others. . . . He is bored with conversation, but greets visitors just like a little dog; yet when the least thing is snatched away from him, he flares up and growls. . . . He is malicious and bites people with his sarcasms.

His preoccupation with introspection and his maladies was so extreme that it gives the impression of a person who lived more through these mirrorings on paper than in reality. The diary entries were a private, secret ritual, and seemed to be a protective shield against the fundamental unhappiness that originated in his fatherless, troubled, and nomadic childhood. The resulting insecurity spurred him on through a tumultuous and intense life, which he perceived as a never-ending series of challenges, questions, and problems.

Such low self-esteem and high vulnerability were concealed by an aggressive and arrogant outward mien: "This man was born destined to spend much time on difficult tasks from which others shrunk," he said of himself. "In philosophy he read texts of Aristotle. . . . He explored various fields of mathematics as if he were the first man to do so. . . . He argued with men of every profession for the profit of his mind . . . [and] defended the opinions of Copernicus."

Although Copernicus's *The Revolutions* of a half century earlier had begun to convince people of a sun-centered universe, the other ancient astronomical beliefs persisted as the world entered the last decade of the sixteenth century: unchanging heavens, perfect circles, celestial harmony, mathematical harmony, uniform motion. Kepler would soon play a major role in uprooting those beliefs. Geometry had a particular significance for

Kepler because, as he wrote, it "existed before the Creation. It is co-eternal with the mind of God. . . . Geometry provided God with a model for the Creation. . . . Geometry is God Himself." This combination of the ancient Greek picture of the universe and the principles of basic geometry formed the cornerstone of Kepler's theory.

Early in his life he wrote that he was convinced that his work was destined to reveal great truths about the universe. This self-proclaimed rendezvous with destiny was the subject of much of his voluminous writings. Yet by any objective standard there was no basis for such a prediction. Other astronomers and mathematicians of his time were more intelligent, better educated, held higher posts, and many certainly had greater opportunities to achieve greatness by virtue of their social status and more likeable personalities. However, on July 9, 1595, while instructing his class in math and astronomy in Gratz, Austria, Kepler was struck with the revelation that he'd discovered the secret of the universe. "The delight that I took in my discovery," he later wrote, "I shall never be able to describe in words." His idea was an elaboration of a concept that had developed over the previous centuries that the universe is built around certain symmetrical geometric figures. He

FIGURE 1-2

Geometric Harmony, from Kepler's *Epitome of Copernican Astronomy*

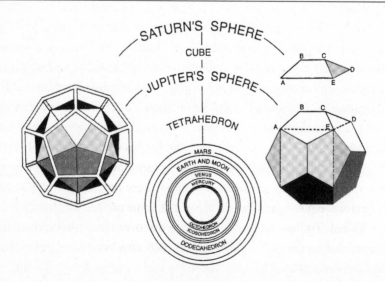

began with two-dimensional shapes. "The triangle is the first figure in geometry," said Kepler. "Immediately I tried to inscribe into the next interval between Jupiter and Mars a square, between Mars and Earth a pentagon, between Earth and Venus a hexagon." He then expanded this theory to the only five solids that can be constructed in a three-dimensional space where each face of the solid is an identical polygon—the so-called perfect solids (tetrahedron, cube, octahedron, dodecahedron, and icosahedron), which he said filled each space between the six planets known at the time (Mercury, Venus, Earth, Mars, Jupiter, and Saturn).

With his revelation about geometric harmony, Kepler was more convinced than ever that he'd been correct about his destiny:

> Since . . . the full sun illuminated my wonderful speculations, nothing holds me back. . . . I dare frankly to confess that I have stolen the golden vessels of the Egyptians to build a tabernacle for my God. . . . The die is cast, and I am writing the book to be read either now or by posterity. . . . It can wait a century for a reader, as God himself has waited six thousand years for a witness.

From that moment forward, Kepler was obsessed with mathematically proving this concept of the solar system—that is, fitting the five perfect solids in the spaces between the six planets. He regarded it as the ultimate insight to God's plan of the universe. He was convinced there was "a *reason* for the number of planets," and began a frantic pursuit for the truth, sustained by a vision of mathematical and geometric harmony in the sky. It was not conceivable to him that the fact there are five perfect solids and six planets was merely a coincidence or that there could actually be more than six planets. His tenacity and commitment to his theory sent him on a wild goose chase spanning many years of his life, for his basic theory was simply wrong.

Kepler went to work for Tycho in 1600. On his deathbed in 1601, Tycho bequeathed Kepler all of his precise and voluminous records of planetary movements. According to Kepler, Tycho lay there repeating over and over, "Let me not seem to have lived in vain. . . . Let me not seem to have lived in vain." Indeed, Tycho's legacy led Kepler to discover three principles that he developed during the period from 1609 to 1618, now known as Kepler's Laws of Planetary Motion:

1. All planets revolve around the sun in elliptical orbits.

2. A radius line joining any planet to the sun sweeps out equal areas in equal lengths of time.

3. The square of the period of revolution of a planet is proportional to the cube of its mean distance from the sun.

With the third law, he gave a precise timetable of the planetary motions, allowing people to calculate the exact location of each planet at any given moment. The second law is depicted in Figure 1-3.

People had spoken and written of an unseen *force* driving the motion of heavenly bodies—commonly conceived of as God. But no one had ever provided such a mechanistic, mathematical interpretation or explanation. In reference to his three laws, he wrote:

Either the *animae motrices* [moving spirit or forces of the planets] are feebler as they are more distant from the Sun or there is only one [force] in the center of all the orbits, that is, the Sun, which impels a body more violently as it is nearer, but which becomes ineffective in the case of the more distant bodies.

FIGURE 1-3
Kepler's Second Law

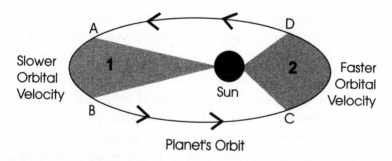

Planet's Orbit

- Time of orbit from A to B equals C to D
- Area 1 equals area 2

In the course of trying to understand God's plan in terms of geometric shapes and harmonies, Kepler stumbled across concepts and mathematical principles that accurately describe the relationship between each planet's distance from the sun and the length of its year. Kepler was the first person in history to tie the planets' motions to their distance from the sun and to think of cosmological problems in terms of *physical forces* rather than divine guidance. Thus astronomy and physics became joined for the first time through Kepler's Laws of Planetary Motion. Ironically, these laws proved that his original theories about geometric harmony were not possible. The perfect circular orbits, which Kepler once considered sacred, did not exist. Instead, there were ellipses (first law). The planets did not revolve uniformly. Instead, their velocity depended on their distance from the sun (second law). And there seemed to be this strange relationship, as if the sun had the power of God. It was devastatingly painful. Indeed, Kepler was actually ashamed of his first law because it contradicted the sacred belief about circular movements, but he had the wisdom and integrity to express his discoveries as he found them, despite their running counter to his grand geometric theory.

Nevertheless, even after he wrote down the three laws, ancient Greek philosophy loomed over Kepler as an obstacle to scientific progress, and he renewed his search for the geometric figures and perfect orbits, almost as if he could not believe his own discoveries. The magical and subjective still tugged at him to go back to the perfect circles and crystalline spheres of Aristotle and Ptolemy. His *Harmonies of the World*, published in 1618, only incidentally includes the third law of planetary motion through which we can calculate the precise timetable of planetary position. But in the same breath, he continued his effort to solve the cosmic mystery with the perfect three-dimensional geometric figures in *combination* with two-dimensional shapes inscribed into the circle. Integral to this grand scheme, Kepler's "harmony" of the universe (in *Harmonies of the World)* is a literal reference to music: "For the extreme diverging intervals of Saturn and Jupiter make slightly more than the octave; and the converging, a mean between the major and minor sixths." The book describes an elaborate mathematical correlation between the planets' movements and musical scales; thus he envisioned a universe that is a synthesis of ancient Greek philosophy, astrology, astronomy, geometry, and music. Yet the three true laws of planetary motion are hidden within his elaborate conceptualization of the universe and persisted for decades among the false axioms in Kepler's writings. Paradoxically, to the day he died Kepler never recognized the significance of these principles, but seventy years later, Isaac Newton would base an important part of his physics on them.

Though Kepler remained convinced of the "harmonies," his brilliant observations combined with a series of mathematical calculations allowed him to carry Copernicus's baton one more step into the modern world. He did so unwittingly, as he straddled two worlds, torn between the old and new, trying to reconcile two schools of thought. Johannes Kepler would be more than astounded today if he could see NASA's *Kepler* spacecraft searching for Earth-like planets orbiting their sun-like stars in distant galaxies. Launched in 2009, the *Kepler* identifies planets by measuring dips in starlight when the suspected planet passes between its home star and the spacecraft. In 2011, planet Kepler-22b was discovered circling a star very much like our own sun. With an annual orbit of 290 days and a likely surface of water and rock, Kepler-22b is one of more than 1,000 "exoplanets" (those outside our solar system) identified by the *Kepler* spacecraft. Whether these planets could be habitable is a topic of growing enthusiasm. Johannes Kepler surely would be thrilled to know that it would take only 22 million years for NASA's space shuttle to reach planet Kepler-22b, located 600 light-years away from Earth.

Over the succeeding decades, it became clear that one must rely on facts and reject unsupported speculation when analyzing the physical world, yet powerful forces persisted in maintaining centuries-old myths and mystic teachings. It was time for courage and thought to coalesce in the hearts and minds of individuals capable of discerning the clear line between astrology and astronomy, between alchemy and chemistry, metaphysics and physics, faith and reason. In Italy at the turn of the sixteenth century there was a growing and profound effort to place both feet on one side of that line, as the irreconcilability of these views became much more than an intellectual dilemma. It became a matter of life and death.

Chapter Two
THE IMMOVABLE EARTH

Much of human history can be described as a gradual and sometimes painful liberation from provincialism, the emerging awareness that there is more to the world than was generally believed by our ancestors.
—Carl Sagan, *Broca's Brain* (1974)

Bruno Is Burned at the Stake as the Church Attempts to Eliminate Threats to Its Authority

A Murder in Rome and a Homecoming in Venice

The chain of events that led to Giordano Bruno (1548–1600) being burned alive at the stake in a Roman marketplace began in 1575 when he secretly read two forbidden commentaries by Desiderius Erasmus (c. 1466–1536), the Dutch humanist and great scholar of the Renaissance. Bruno's impetuousness and his disdain for dogmatic restraint made his fate inevitable, but reading the writings of Erasmus that questioned some church teachings labeled Bruno a heretic and hastened that fate, as he fought for freedom of thought and expression for the remainder of his life.

In July 1575, Bruno, who'd been ordained as a priest three years earlier, completed the prescribed course in theology at the Dominican Convent of San Dominica Maggiore in Naples. Up to that point, he hadn't publicly criticized the Dominican Order or the church, but his unorthodox theological views were known, and his enemies reported that he'd read those censored works of Erasmus. This was not a matter of Bruno himself writing or advocating disobedience to the church in those early years of his thinking as a theologian and philosopher. Yet in the paranoid and tortuous rationale by which the church attempted to control thought and morality, the mere consideration or discussion of Erasmus's objectionable views was sufficient for a trial for heresy to be prepared against Bruno, forcing him to flee Naples for Rome in February 1576. He was unjustly accused of murder there and also found himself the subject of a second excommunication process. By April 1576, he was again a fugitive, this time settling in Geneva, where he was arrested and excommunicated on the basis of an article he wrote criticizing a Calvinist professor. Nevertheless, Bruno was soon allowed to retract the article—thus rehabilitated, he was permitted to leave the city.

Between 1580 and 1585, he became a well-known writer in the fields of theology, astronomy, and philosophy and was a professor in Paris and London, and at Oxford University, where new thought was welcome. During those years, he wrote extensively on the Copernican theory of the solar system and suggested that the universe was infinite. He said the Bible should be followed for its moral teachings but not its astronomical statements. He rejected Aristotelian physics and criticized the Calvinistic principle of salvation by faith alone.

Beginning in 1585, political developments narrowed the margin of tolerance in which Bruno had lived and prospered for the previous five years. Yet almost as a challenge to those in power in Paris, in 1586 Bruno wrote a series of articles in which he insulted a high government official and renewed his attack on Aristotle. For this he was again forced out, leaving Paris for Germany where he lectured at various universities and published several articles. He closed out the decade by being excommunicated once again, this time in Helmstedt, Germany, by the local Lutheran Church.

Then, in August 1591, fifteen years after leaving his homeland, he came full circle and returned to Italy, a decision that would ultimately prove fatal, but one in keeping with his tenacious devotion to his principles. When the Venetian patrician Giovanni Mocenigo invited him to return, Bruno viewed it as a safe and reasonable decision. Venice in 1591 was the most liberal of the Italian states and appeared to be a haven for unorthodox views. Religious tension in Italy and its neighbors had eased after the death of the uncompromising Pope Sixtus V in 1590, and there seemed to be a trend toward religious pacification. Another incentive for returning to Italy was Bruno's hope he'd be appointed to fill the vacancy in the prestigious mathematics chair at the University of Padua. So he began lecturing and writing at Padua in the late summer of 1591, but returned to nearby Venice in early 1592 when it became apparent the chair would be offered to Galileo.

In the spring of 1592, there was a falling out between Bruno and his host Mocenigo, who turned on him unexpectedly and maliciously by denouncing him to the Venetian Inquisition for heresy. In the course of this arrest and trial, Bruno had another opportunity to justify his actions, retract his philosophy, and go on with his life. He admitted errors and pointed out that his theories were philosophy, not theology, thus not intended to question the church's power. Yet, for higher authorities the memories of Bruno seventeen years earlier were still vivid, and the Roman Inquisition demanded his extradition. On January 27, 1593, Bruno became a prisoner of the Holy Office of the Roman palace and began a trial that was to last *seven years*. The charges against him were based primarily on his writings in which he stated that the Earth is not at the center of the universe, the universe is infinite, and the stars are not fixed on a crystal sphere.

The inquisitors informed him that only an unconditional retraction of all his theories would save him. Faced with the choice of uttering lies to indulge his intolerant accusers or maintaining adherence to his principles and losing his life, he declared he had nothing to retract. Pope Clement VIII ordered the death sentence on February 8, 1600, and nine days later, at the Campo de' Fiori, Giordano Bruno, bound and his tongue in a gag, was set afire—transforming him from merely a progressive thinker to a martyr for freedom of thought and expression. Bruno's writings later influenced Galileo and became an important source of scientific thought for centuries after his death.

Galileo's Improved Telescope Reveals the Heavens

A Dispute Breaks Out . . . Jupiter's Moons

Born in Pisa, Italy, on February 15, 1564 (two months before Shakespeare's birth), Galileo Galilei was the eldest of seven children and was raised in a household that valued the arts and welcomed new ideas. His father, Vincenzio, was a cloth merchant by trade, and an accomplished musician and composer. In 1574, he moved the family to Florence, where Galileo was sent to the famous Jesuit Monastery school at Vallombrosa when he was twelve. He enrolled at the University of Pisa at seventeen as a medical student and quickly developed a reputation for obstinately disputing the nonsensical doctrines handed down from Aristotle and the Greek physician Galen (129–199 CE).

Galileo eventually abandoned his medical studies to pursue mathematics, mechanics, and hydrostatics. His rapid mastery of those subjects can be largely attributed to his mentor and tutor at the University of Pisa, Ostilio Ricci, whose teachings emphasized practical application of math principles, which ideally fit Galileo's developing view of the universe and his prowess as an inventor. In his first year at the university, his attention was drawn to a swinging lamp and the fact it always seemed to require the same amount of time to complete an oscillation, regardless of the range of the swing. After Galileo verified his observation through experiments, he suggested that the constant regularity of the pendulum could be used to construct a clock that would be highly accurate. He also applied the principle in an invention for measuring the pulse.

Forced to leave school in 1585 for lack of funds, Galileo continued studying on his own and sufficiently developed his knowledge of physics to be appointed a lecturer at the Florentine Academy when he returned to Florence. In 1586, he published an essay describing the invention of the hydrostatic balance (a scale allowing for precise measurement), which brought him to prominence throughout Italy. However, he was still unable to find a steady source of income. But in 1589, at age twenty-five, primarily as a result of a treatise he wrote in 1587 on the center of gravity in solids, Galileo earned the chair of mathematics at the University of Pisa, where he wrote and taught on a variety of subjects, including the laws of motion, for the following two years.

It was at Pisa that Galileo was said to have dropped two balls of unequal weight from the Leaning Tower to prove that they would both fall and accelerate at the same rate, contrary to the writings of Aristotle, who said the heavy ball would reach the ground before the light one. Historians are generally convinced that Galileo never performed that experiment. They surmise that he either conducted a demonstration at the Leaning Tower (of a result he already knew) or he performed a "thought experiment" like the following to prove the point: Imagine two unequal balls dropped from the Tower at the same time, said Galileo, and also suppose that Aristotle was correct and the heavy ball would fall faster. But now imagine the same experiment with one difference—namely, that the two unequal balls are joined by a string or cable between them. If it were true that the heavy ball moves faster and the light one slower, then the light one will hold back the heavy one. If Aristotle were correct, the two balls tied together would not reach the ground as quickly as the heavy ball alone. But if we assume that the string or cable between the balls has the effect of turning the two balls into a single mass, which is *heavier* than either one by itself, the tied balls should drop faster than either one by itself. In fact, in a vacuum, where air resistance is not a factor, all objects fall at the same rate. A feather falls as fast as a cannonball.

Galileo was forced to resign from the University of Pisa in the summer of 1592. Historians have various theories concerning his departure. Some maintain that his lectures on his discoveries concerning the motion of falling bodies might have offended and alienated the members of the faculty who still believed in the teachings of Aristotle and who saw their vested interests being threatened by Galileo. Also, he was highly critical of the university's regulations. In another theory, Professor Stillman Drake (translator and biographer of Galileo), suggests that political pressure was used against Galileo by a son of the Grand Duke of Tuscany whom he'd offended. Whatever the reason, Galileo left Pisa and his native Tuscany to fill the mathematics chair at the University of Padua in the Republic of Venice, having been selected over Giordano Bruno (sixteen years Galileo's senior) for the prestigious position. Galileo remained at Padua for eighteen years (1592-1610), teaching math and astronomy, and at the same time accomplishing much of the work that established his reputation as a scientist and inventor and that was later used by Isaac Newton to form the foundation of modern physics. During the years at Padua, he lived with his mistress Marina Gamba and fathered two daughters (in 1600 and 1601) and a son (in 1606).

Galileo's scientific contribution as the pioneer of modern physics was more significant than his work in astronomy or as an inventor. His research and discoveries in physics, which would later be used by Newton, were set forth in these writings:

- 1585-1587: Writes essays on motion and centers of gravity of certain objects.

- 1588: Writes landmark paper on motion, "*De motu.*"

- 1589: Writes essay on logic titled "Demonstrations in Science."

- 1591: Conceives axial rotation of Earth.

- 1593: Composes summary on branch of physics to be known as mechanics.

- 1595: Explains ocean tides in terms of Copernican motion of Earth.

- 1601: Completes *The World Systems*, and analyzes Kepler's data.

- 1602: Begins studies on magnetism and pendulum motion.

- 1603–1604: Develops theorems on motion and falling objects.

- 1612: Publishes book on the motion of objects in water.

- 1623: Writes "The Assayer," on the scientific method.

- 1624: Begins *Dialogue on the Tides*.

- 1625–1631: Further writing on motion, tides, and falling bodies.

- 1632: Publishes *Dialogue on the Two Chief World Systems*.

- 1638: Publishes *Two New Sciences*.

Of all his discoveries and inventions, his most famous is the telescope. Yet, as with the legend of dropping the balls from the Leaning Tower of Pisa, there is also an element of myth in the story about his telescope. Galileo himself wrote, "We are certain the first inventor of the telescope was a simple spectacle-maker who, handling by chance different forms of glasses, looked, also by chance, through two of them, one convex and the other concave, held at different distances from the eye; saw and noted the unexpected results; and thus found the instrument." Credit for the invention is generally given to a Dutch spectacle maker named Hans Lippershey who tried to sell the telescope to the government of Holland in 1608 as a military instrument for its war against Spain. After a special committee recommended the purchase, a dispute broke out in Europe among several people, each claiming he was the inventor. Galileo wasn't among them.

By the end of 1608, telescopes were being made and sold throughout Europe for navigation, military uses, and astronomy, and as novelty items. When the Venetian senate expressed a desire to buy the new seeing device for maritime military uses, a government official aware of Galileo's expertise in instrument making asked him to make a telescope of his own design. In July 1609, Galileo began experimenting with ways to make the lenses, and in just one month he'd made an instrument three times more powerful than any in existence, and presented one to the senate as a gift. By the end of the year, he'd made a thirty-power telescope, more than tripling his previous efforts. It was this superior talent as an instrument maker that resulted in Galileo receiving the primary credit for *inventing* the telescope. His decision to quickly turn the device upward to the heavens resulted in his fame as an astronomer and further enhanced the historical link between Galileo and the telescope. That decision also brought about the tumultuous events that would dominate the rest of his life.

"Our sense of sight," Galileo wrote in March, 1610, after looking through the telescope, "presents to us four satellites circling about Jupiter, like the Moon about the Earth, while the whole system travels over a mighty orbit about the Sun." This is from his pamphlet *"Sidereus Nuncius"* (The Starry Messenger), which was the first written account of observations of celestial objects through a telescope. In addition to Jupiter's moons, Galileo described the mountainous surface of our own moon, and explained that the Milky Way consists of stars rather than a white cloudy substance, as previously thought. This work with the telescope immediately led to his appointment as philosopher and "mathematician extraordinaire" to the Grand Duke of Tuscany. He left his position at the University of Padua to fill this lucrative post and devote more time to research. Praise came from all quarters. He was invited to Rome and honored by a meeting with the pope on April 1, 1611, during which he gave a demonstration of the telescope, and on April 14, 1611, the Academy of Lynxes (a scientific society in Rome) held a banquet in Galileo's honor. For several years, Galileo enjoyed the freedom to report his observations publicly and without concern for how they might conflict with the entrenched beliefs of the day. He had no reason to suspect that these observations would soon become seeds of controversy and tragedy.

Galileo Begins to Publicly Support a Sun-Centered Universe as the Church Issues a Warning

Letters to the Duchess on Sunspots . . .
God Stops the Sun for the Israelites

As early as April 4, 1597, Galileo had written to Kepler that he'd "become a convert to the opinions of Copernicus many years ago," yet he continued to teach the Ptolemaic system throughout his eighteen years at Padua because he felt more evidence was needed. But in 1610, after his own observations with the telescope, Copernicanism became a clearly supportable and objective truth for him, and he hinted at these views in "The Starry Messenger," though stopped short of fully endorsing Copernicanism. His discoveries of the moons of Jupiter, the rings of Saturn, and the phases of Venus revealed the reality of this solar system in a way that Copernicus himself could not have known.

The church advised Galileo that it disagreed with the interpretations and descriptions in *The Starry Messenger* but did not interfere with his right to express his views or continue those observations. Then began the conflict that would dominate Galileo's remaining years. In 1613, stepping into the unknown chasm of new ground broken by Bruno, Galileo began publicly to support a heliocentric solar system. This is a new theory, Galileo later said, on which "all my life and being henceforth depends." The flames that engulfed Giordano Bruno in 1600 had since burned out, but the political climate of Italy that had ignited the fire was not very different in 1613. Dancing on the edge of the church's favor and license, Galileo's teachings, writings, and theories were no longer calmly accepted. As noted by J. L. Heilbron in his 2011 book, Galileo had brought "some fundamental problems in the culture of his time so crisply into conflict that they could not be avoided or resolved."

In December 1613, Madame Christina of Lorraine, Grand Duchess of Tuscany (mother of the grand duke) had engaged Galileo's friend and student Benedetto Castelli, a mathematician and Benedictine abbot, in a discussion over dinner at the royal court. Numerous dignitaries were present, including the grand duke, but Galileo wasn't there. Castelli found himself under siege and having to defend Galileo's discoveries against the traditional "wisdom" of the group. After Galileo received a letter from Castelli describing the event, he composed the first of a series of letters that came to be called the *Letters on the Sunspots*, discussing the relationship between science and religion, and sent it to Castelli in the hope that it would be useful if he ever found himself in a similar predicament. Following Bruno's ill-fated lead, Galileo wrote that the Bible should be followed for its moral teachings but that it does not contain the answers to the mysteries of nature.

During most of 1614, opposition to Galileo and his public views was quietly forming among jealous university colleagues and within the Catholic Church. The first public ecclesiastical attack was launched from the pulpit of Santa Maria Novella in Florence on December 21, 1614, when Father Thomas Caccini denounced Galileo. Caccini first referred to the passage in the Bible when Joshua beseeched God to halt the sun so the Israelites would have sufficient daylight to sustain their momentum and defeat the Amorites. Based on this, Caccini posed the question: "If God stopped the sun, how could it be that the sun wasn't moving around the Earth in the first place?" In a vicious and lengthy condemnation, the priest didn't stop at the Copernican system, but indicted Galileo personally as well as mathematics and *all mathematicians* as religious and political heretics.

In February 1615, an influential Dominican priest named Niccolò Lorini was given a copy of Galileo's private letter to Castelli, and reported Galileo's "heretical views" to the holy office in Rome. Galileo heard of Lorini's action and, wary of the problems the Castelli letter could cause, prepared a modified version of it and sent it to his friend Piero Dini in Rome and asked him to show it to Cardinal Robert Bellarmine, the church's chief theologian, along with a cover letter, dated February 16, 1615, downplaying some of the points in the original version that were in conflict with the scriptures. However, in a letter dated March 12, 1615, Castelli informed Galileo that the archbishop of Pisa had demanded that Castelli relinquish the original letter to him and that the archbishop had said "it was soon to be made known to you, Galileo, . . . that these ideas are all silly and that they deserve condemnation."

The battle lines were becoming clear. While the opposition coalesced there was also a loyal and growing band of supporters, including a number of Jesuits. One priest, Paolo Antonio Foscarini, had even written a book devoted to defending the Copernican system from charges that it was inconsistent with the Bible and had sent a copy to Cardinal Bellarmine for his opinion shortly after the cardinal had received the revised version of the letter to Castelli. However, Bellarmine resoundingly rejected Foscarini's book and inferred grave consequences for those who supported the views of Copernicus and Galileo.

Thus, amid this storm of controversy brewing in the spring of 1615, Galileo was faced with the choice of abandoning his position altogether or demonstrating that the Bible can be reconciled with Copernican theory, for he had never actually believed the two to be inconsistent despite the allegations of his detractors. He decided to defend himself in a reasoned and cautious manner. God's truth, he wrote, is communicated in two forms—the Bible and nature. "None of the physical effects that are . . . placed before our eyes . . . should ever . . . be placed in doubt by passages of the Scriptures which seemed to have a different verbal import. . . . Two truths can never contradict one another." The primary vehicle for supporting his position was an amended and expanded version of the views originally expressed in Galileo's letter to his friend Benedetto Castelli. In its new form, it became Galileo's famous "Letter to Madame Christina of Lorraine" (subtitled "Concerning the Use of Biblical Quotations in Matters of Science"), completed in June 1615.

> Some years ago . . . I discovered in the heavens many things that had not been seen before our own age. The novelty of these things, as well as some consequences which followed from them in contradiction to the physical notions commonly held among academic philosophers, stirred up against me no small number of professors—as if I had placed these in the sky with my own hands in order to upset nature and overturn the sciences. . . . Showing a greater fondness for their own opinions than for truth, they sought to deny and disprove the new things which, if they had cared to look for themselves, their own senses would have demonstrated to them. To this end they hurled various charges and published numerous writings filled with vain arguments and they . . . cast against me imputations of crimes which must be, and are, more abhorrent to me than death itself. . . .

I hold the sun to be situated motionless in the center of the revolution of the celestial orb while the earth rotates on its axis and revolves
about the sun. They . . . resolved to fabricate a shield for their fallacies out
of the mantle of pretended religion and authority of the Bible. . . . They
had no trouble in finding men who would preach the damnability and
heresy of the new doctrine from their very pulpits.

The battle was on for the hearts and minds of all Europeans. But the
aristocracy and church powers were not ready for a fundamental change in
thinking about the universe or, more important, to relinquish any aspect of
authority, including the church's claim to be sole interpreter of the Bible.
Science as an institution was new and precarious, particularly physics and
astronomy. Religion and the authority of the church, on the other hand, were
firmly embedded in the minds and culture of the day. In his letter to Madame
Christina, Galileo appealed to logic over emotion:

I do not feel obliged to believe that the same God who has endowed us
with senses, reason, and intellect has intended to forego their use and by
some other means to give us knowledge which we can attain by them. . . .
The intention of the Holy Ghost is to teach us how one goes to heaven,
not how heaven goes.

Before embarking on a lengthy series of examples showing that the Bible
cannot be interpreted literally and that much of it is subject to differing interpretations, Galileo warned of the grave consequences for society and its
citizens if no one is allowed to suggest such differing interpretations. "Who
indeed will set bounds to human ingenuity?" wrote Galileo. "Who will as-
sert that everything in the universe capable of being perceived is already
discovered and known?"

However, it didn't matter whether Galileo's reasoning in the letter to
Madame Christina was compelling or whether his evidence of the physical
movement of the planets was true and comprehensible by the average person. His effort to rally support for his views was failing because the rise of
modern science remained weighted down by the 2,000-year history and system of erroneous beliefs that had been embraced by the vast majority of
people. Also, as the Roman Catholic Church continued its fierce attack

against the Protestant Reformation, which had been so successful during the 1500s, and because there was a broader struggle for authority among the secular forces, the Catholic Church, and the Protestants throughout Europe, Galileo's views gradually became more a symbol for being on one side or the other of a political position than they were a developing area of astronomy and mathematics. In other words, this was more an issue of individual freedom to express views than the accuracy of the Copernican system. Many of the details of Galileo's proof were far too complex for those who chose not to listen, but even when he invited the Jesuit fathers to look through the telescope and observe Jupiter's moons themselves, they refused. Instead, they continued to express their dissatisfaction with Galileo's inability to "strictly demonstrate" the truth of his position when it is so clear to every human that the Earth stands still.

Galileo Is Tried for Continuing to Defend Copernicanism

The Decree Is Broken . . . Even So, the Earth Does Move

Against the advice of his friends, Galileo insisted on traveling to Rome in December 1615 to further defend his work. Convinced that Galileo's views could potentially undermine the Catholic Church's fight against Protestantism, on March 5, 1616, Cardinal Bellarmine decreed that Copernicanism is "false and erroneous" and banned the writings of Copernicus, stating that "God fixed the Earth upon its foundation, not to be moved forever." At about the same time, Cardinal Bellarmine warned Galileo not to hold or defend the Copernican doctrine, although it could be discussed as a mere "mathematical supposition." Several years then passed without incident, as Galileo led a quiet life in his house in Bellosguardo near Florence.

However, in 1620 the church published its "corrections" to Copernicus's *The Revolutions*, and in 1622 the church created the Institution for the Propagation of the Faith (from which the word *propaganda* is derived). Despite the warning signs that Galileo must have seen from these developments, he went to Rome in 1624, hoping to secure a revocation of the Decree of 1616. Although unsuccessful in this, he did obtain permission from his friend, the new Pope Urban VIII (a Barberini), to write about "the systems of the world," which Galileo interpreted as permission to publish a fair dialogue examining and comparing the Ptolemaic and Copernican views. Attempting to work within this narrow grant of authority, Galileo returned to Florence and spent the next several years working on his most famous book, *Dialogue on the Two Chief World Systems*. Published in Florence on February 21, 1632, and greeted with praise from scholars across the European continent as a literary and philosophical masterpiece, this book would ultimately destroy Galileo.

Shortly after the book's publication, Galileo's now-estranged friend and former supporter, Pope Urban VIII, established a special papal commission to investigate his writings. Based on the commissions' report, Galileo was summoned before the commissary-general of the Holy Office in Rome and tried by the ten judges of the Roman Inquisition for espousing the Copernican theory of the structure of the universe, thereby violating the Decree of 1616. He was not allowed to have a copy of the charges or the evidence and he had no counsel to defend him.

Galileo, the most celebrated scientist in Europe at that time, placed his faith in the demonstration of his principles and asked the church officials to suspend the dogma and authority of the Scriptures on which they had based the indictment. But despite his age, his poor health, and, indeed, the clear truth of his spoken and written words, this verdict was reached: Galileo's views, like those of Copernicus, were judged "false and erroneous," and he was ordered to retract his words publicly and never teach Copernican theory again. Thus, thirty-three years after Giordano Bruno was burned alive, Galileo faced the same choice that had been offered to Bruno. In a decision that some have criticized as damaging the cause of science, Galileo chose life. On June 22, 1633, he gave the lengthy recantation that read in part:

> I, Galileo Galilei ... aged seventy years ... and kneeling before you, most Eminent and Reverend Lord Cardinals ... swear that I have always believed, do now believe, and by God's help will for the future believe, all that is held, preached, and taught by the Holy Catholic and Apostolic Roman Church. ... Whereas, after an injunction had been judicially intimated to me by this Holy Office ... that I must altogether abandon the false opinion that the sun is the center of the world and immovable, and

that the earth is not the center of the world, and moves, and that I must not hold, defend or teach . . . the said doctrine. . . . I wrote and printed a book in which I discuss this doctrine . . . and for this cause I have been pronounced by the Holy Office to be vehemently suspected of heresy. . . . Therefore, desiring to remove . . . this strong suspicion . . . I abjure, curse and detest the aforesaid errors and heresies, and generally every other error and sect whatsoever contrary to the Holy Church. . . . I . . . have . . . sworn, promised and bound myself as above.

Legend has it that as Galileo rose from his knees, he murmured, "*E pur, si muove*" (Even so, it does move). To ensure he would comply with the sentence, he was shown the instruments of torture. He was confined in his villa for the remaining days of his life under strict house arrest. During those years, he wrote his last and scientifically most important work, *Two New Sciences*. Because the Inquisition forbade his printing any book, the manuscript was smuggled to France for publication. In 1638, Galileo became blind before he was able to receive a copy of this book. Still a prisoner in his own house, he died January 9, 1642.

It wasn't until 1757 that the church removed Galileo's *Dialogue on the Two Chief World Systems* from its list of banned publications. In 1979, Pope John Paul II asked the clergy to reconsider the absurd claim of heresy and to "close the Galileo affair." On October 31, 1992, 359 years after the Roman Catholic Church condemned Galileo and forced him to recant the truth, the Vatican formally acknowledged its error in a statement delivered by Pope John Paul II to the Pontifical Academy of Sciences. "The theologians who condemned Galileo," said the Pope, "did not recognize the formal distinction between the Bible and its interpretation. This led them unduly to transpose into . . . faith a question which in fact pertained to scientific investigation." In 2009, on the eve of the 400th anniversary of Galileo's initial observations with the telescope, the Vatican launched a new approach to atone for the way he was treated. It erected a statue of Galileo in its garden. Pope Benedict XVI stated that Galileo and other scientists like Copernicus and Tycho Brahe were among those who have explained the laws of nature and that 2009 was the International Year of Astronomy. In May 2009, representatives from the Vatican Observatory and the Pontifical Academy of Sciences, among others, gathered in Florence to again examine the circumstances that led the Catholic Church to accuse Galileo of heresy. Despite this belated admission, the dispute between Galileo and the church still stands as one of history's great symbols of conflict between reason and dogma, science and faith.

During the hundred years spanning the lives of Tycho, Kepler, and Galileo, Aristotle's universe of crystal spheres on circular paths became a wistful image, Ptolemy's epicycles and geocentric theory of the universe was disproved, and in the ensuing decades the Roman Catholic Church lost its control of government and the people. These three individuals were responsible for laying the foundation for the rise of astronomy and the beginning of physics in the seventeenth century. The scientific method of inquiry and its criterion of truth and the modern idea of experiment can be attributed primarily to Galileo. His improvement of the telescope, his resultant observations and writings, his tying theory and experiment to one another, and his use of mathematics in physics all contributed to the remarkable scientific discoveries that were to follow. His life and his work would touch another great figure—later in 1642, the year Galileo died, Isaac Newton was born. Newton would lift the discoveries of Galileo and his other predecessors to an unprecedented height. He would vindicate them all and would incorporate their efforts into the first of the seven greatest scientific discoveries in history.

Chapter Three
THE PRINCIPIA

It appeared to him reasonable to conclude that this power must extend much farther than was usually thought; why not as high as the moon . . . ? He considered . . . if the moon be retained in her orbit by the force of gravity, no doubt the primary planets are carried round the sun by the like power.
—Henry Pemberton, *A View of Sir Isaac Newton's Philosophy* (1728)

I do not know what I may appear to the world; but to myself I seem to have been only like a boy playing on the seashore, and diverting myself in now and then finding a smoother pebble or a prettier shell than ordinary, while the great ocean of truth lay all undiscovered before me.
—Isaac Newton (1727)

The Academy for Experiment was founded in Italy in 1657 and the Royal Society of London was founded in 1660. The first scientific articles were published in England and France a few years later, as Europe gradually passed from old methods and approaches to the modern ways of scientific thought. Thus Isaac Newton, born in 1642, grew up in the midst of a growing intellectual movement of which he was to become the leading figure at an early age.

Isolated for Eighteen Months, Newton Discovers the Laws of Physics, but Does Not Publish for More Than Twenty Years

The Apple and the Moon

Newton was born on Christmas Day near the village of Colsterworth in Lincolnshire County, England, in a small stone farmhouse that still stands in the tiny hamlet of Woolsthorpe. His mother later said the baby was "so little they could put him into a quart pot." Born several weeks premature, he was weak and sickly and wasn't expected to survive but soon overcame the odds and predictions. His father (also named Isaac) had died three months

FIGURE 3-1
Newton's Tree

Author David Brody and his wife, Susan, standing in front of the apple tree in the garden next to Isaac Newton's childhood home, Woolsthorpe, England. The original tree was destroyed in a storm in 1820 and this tree grew from the roots of the original.

before, and left the 100-acre farm and the modest two-story house to his wife, Hannah. In 1642, a garden grew to the west of the house and included an apple tree that was to loom large in history.

A month after Newton's third birthday, his mother married Barnabus Smith, a reverend in North Witham, a town a mile from Woolsthorpe. For reasons that young Isaac would never understand and are still not clear to historians, his mother moved to Smith's home and left her son in her mother's care. Over the ensuing years Newton and his mother seldom saw each other, as she became preoccupied with raising the three children born to her and Smith. A few years later, Isaac was old enough to climb trees at Woolsthorpe and see North Witham and the spire of the Reverend Smith's church in the distance, reminding him of this traumatic loss for the next several years, until Smith died and Newton's mother finally returned to Woolsthorpe in 1653, when Newton was ten years old. During his teenage years, Newton occasionally wrote about this agonizing period. For example, he recalled that he'd felt like "threatening my father and mother Smith to burne them and the house over them." The abandonment, some thought, could explain his sensitive temperament that was often displayed later in life.

At age twelve, he enrolled in King's School in Grantham, seven miles from his home, and distinguished himself as an outstanding model maker, but didn't yet excel at math or science. During these years he was fascinated with forces generated by moving air and rushing water, often visiting windmills and building scale models of them. He also carefully tracked the sun's movements and constructed sundials and a water clock. Like Leonardo da Vinci and Benjamin Franklin, Newton displayed outstanding mechanical aptitude combined with an innate understanding of geometry, yet he was still a teenager.

At age eighteen he enrolled in the famous Trinity College at Cambridge University where he later met the gifted mathematician Isaac Barrow, who encouraged Newton to study mathematics and optics. Newton received his degree from Cambridge in 1665, but was forced to return to his small home in Woolsthorpe later that year because the Great Plague of London was sweeping through the city. It eventually claimed 75,000 victims, 16 percent of the estimated population of London, causing the university to close for fear of the disease spreading there. This wave of bubonic plague had begun in late fall of 1664 in a London suburb and reached its peak in the middle of 1665, killing more than 68,000 people in London that single year. King Charles II and his court fled from London in June, and the British Parliament was forced to move to Oxford for a short session. In 1666, the mortality dropped to 2,000, but Cambridge University did not reopen until the spring of 1667.

During the "plague years"—the eighteen-month period Newton spent at home, beginning when he was only twenty-three years old—he laid down the foundation of the work that was to revolutionize science. The origin of the famous apple story can be traced to this period. On May 16, 1666, in the course of attempting to conceive of the physical laws that would explain how the moon revolved around the Earth, Newton happened to be sitting near the apple tree in the garden at Woolsthorpe when he saw an apple drop to the ground. At that moment, he realized that the same central pull of the Earth applied to both objects, the apple and the moon, and that

- If the moon were at rest, like the apple on a tree, it would also fall to the Earth.

- It is the Earth's pull—though weakened by the great distance between the moon and Earth—that keeps the moon from flying out of its orbit.

These realizations led him to develop the law and mathematical formulas that account for the pull of the Earth decreasing as objects are farther away from it. He figured out that such pull weakens inversely as the square of the distance from the center of the Earth. To compare the force at two different distances you would square each of those distances and interchange them because the force corresponding to the greater distance is smaller.

So, for example, if one object (such as a planet) were two times farther from the sun than another planet of equal mass, the relative force of gravity would be one over two squared, or one-quarter. In other words, the sun and the more distant planet would pull on each other at one-quarter the force of the sun and the closer planet. At three times the distance, the force would be one over three squared, or one-ninth. And the planets and heav-

FIGURE 3-2

Newton's Inverse Law of Gravity

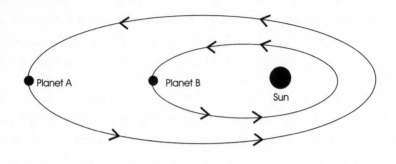

Gravitational force between Planet A and Sun is one-fourth of the gravitational force between Planet B and Sun

enly bodies, like every object, have a mass that is the quantity of matter in that object. The quantity of mass reflects the actual total number of protons and neutrons in the atoms that make up a planet or object, as we will explore in Chapter 5. Unlike the weight of an object, mass doesn't depend on the gravitational pull of the Earth or other body. Therefore, physics uses this universal measure, which is unchanging.

It was during this period at Woolsthorpe that Newton also developed the laws of motion:

1. A body at rest will continue at rest unless a force acts on it, and a body that is moving steadily in a straight line will continue to move with the same speed in a straight line unless a force acts on it. For example, if you place a ball on a level surface, it won't move without some external force acting on it. If such a force causes it to begin rolling, the ball would continue to roll in that direction forever if it weren't for the friction with the surface or an obstacle in its path.

2. Acceleration (rate of change of motion) is directly proportional to the force. For example, the more force generated by an automobile engine, the faster the car will accelerate. Twice as much force will double the acceleration.

3. Every action has an equal and opposite reaction. For example, the action of a bullet being fired by a gun results in the reaction of the gun as recoil.

With these laws of motion Newton established the science of mechanics and laid the groundwork for what is now called classical physics. These principles, including the inverse law of gravity (by which pull weakens inversely as the square of the distance from the center of the Earth) might seem obvious and simple to today's physicists, but this was a new way of thinking in Newton's time—to find simple, precise mathematical laws from which the observed measurements could be worked out in detail. He was able to rely on Kepler's and Galileo's work to some extent in the development of the inverse law of gravity and the laws of motion, which Newton acknowledged when he said that he "stood on the shoulders of giants." But the principle of universal gravitation and its mathematical formula in particular was a monumental advance beyond Galileo's astronomical discoveries and beyond Kepler's Laws of Planetary Motion, which told us only that the planets move in certain regular orbits related to their distance from the sun. Although Kepler touched on a "central force," his laws do not provide insight into the dynamics of the solar system or how such a structure as the solar system can function or exist at all. Newton's idea, on the other hand, was that there exists an invisible force that exerts control over matter without having direct physical contact. This was a bold idea in 1666. It demonstrated for the first time the mathematical formulas supporting the concept Copernicus had touched on—how the solar system can be understood as a totally interrelated dynamical structure.

Physically and intellectually isolated during the Woolsthorpe period, Isaac Newton pondered the motions of the heavenly bodies. "I keep the subject constantly before me and wait til the first dawnings open little by little to the full light," he later wrote. Newton also developed a new branch of mathematics known as differential and integral calculus and began his work in optics. "All this was in the two plague years of 1665 and 1666, for in those days I was in the prime of my age for invention and minded mathematics and philosophy [meaning science] more than at any time since." In those eighteen months, Newton mastered the basic laws of mechanics, convinced himself they applied to all celestial bodies, discovered the fundamental law of gravity, invented calculus, and made significant advances in the field of optics. Nevertheless, he remained very reluctant to publish any of his discoveries because he felt that further experimentation and proof were needed. Certain calculations didn't seem accurate to Newton because he'd relied on the generally accepted (but incorrect) value for the diameter of the Earth. The world would have to wait while Newton worked out this and other problems. He became preoccupied with his lengthy correspondence with several other scientists on color, light and optics, telescopes, grinding lenses, and other subjects. He put his notes aside and temporarily turned away from science to engage in other pursuits, charging himself with the safekeeping of these ideas for over twenty years.

Newton Reveals the Laws of the Universe

A Hard Sell in Latin

Newton returned to Cambridge when it reopened in 1667 and was recognized as a senior fellow of Trinity College. The next year, his mentor, Isaac Barrow, resigned his position as Lucasian Professor of Mathematics and arranged for Newton, only twenty-five years old at the time, to succeed him in that prestigious chair. This is the same position held by the famous theoretical physicist Stephen Hawking from 1979 until his retirement in 2009.

By 1684, Newton still hadn't published any of the principles he'd developed at Woolsthorpe. However, in that year he explained his calculations on planetary motion and elliptical orbits to his friend and colleague, the great astronomer and mathematician Edmond Halley (1656–1742), who determined the orbit of the comet that now bears his name. Halley immediately recognized the overwhelming importance of Newton's work and convinced him to publish his discoveries. At first, Newton could not even find his notes. After finally locating them, he began to organize this material into a cohesive body of theories and mathematical demonstrations. Halley solicited financial support from the Royal Society to have the work published, but when the society backed out, Halley paid the costs of publication himself, negotiated with the printers, edited Newton's writing, wrote the preface, and handled all the details in getting Newton's work published. Thus the world has Halley to thank for the appearance, more than twenty years after Woolsthorpe, of what is now considered the greatest scientific book ever written: *The Mathematical Principles of Natural Philosophy*, commonly known as the *Principia*.

The *Principia* is divided into three "books," although it was published as one volume when it first appeared in 1687. The first book is about the branch of physics called mechanics, and explains the reasons bodies move in the manner they do in empty space. In the second book, Newton deals with the motion of bodies in surroundings that offer resistance, such as air or water. In the third book, using the principles established in the first two books, Newton demonstrates the structure and operation of the entire solar system. He coins the word *gravity* from the Latin word *gravitas*, meaning "heaviness" or "weight." He so accurately explains the motions of the moons of Jupiter, Saturn, and Earth as well as the movements of all the planets around the sun that few significant improvements were made to his work for 200 years thereafter. He then describes how to calculate the masses of the sun and planets from the mass of the Earth, which he also demonstrates mathematically. He explains the precise motions of the moon and that the tides of the seas are due to the gravitational pull of the moon and the sun on those waters. He also calculates the sun's attraction on comets.

The *Principia* is generally considered to be the most prodigious accomplishment of human intellect in the history of humankind and without question the most important single work ever developed in physics. The book immediately brought great fame to Newton. In 1687, the French scientific periodical *Journal des Savants* (Journal of the Learned) stated: "The work of Mr. Newton is the most perfect treatise on mechanics that can be imagined, it not being possible to provide more precise or more exact demonstrations than those which he gives." A twelve-page review of the *Principia* in the German magazine *Acta Eruditorium* (The Transactions of the Learned) in 1687 called Newton the foremost mathematician of the time. The demand for the *Principia* grew steadily in succeeding years. It was published in several other countries, though usually in the original Latin version. It was revised by Newton for a second edition in 1713 and a third edition in 1726, which was the last edition in his lifetime. The book was translated into English for the first time shortly after Newton's death in 1727, and has since been translated into many other languages.

Despite being recognized as a great work, the *Principia* was so lengthy, complex, and difficult to grasp (particularly in Latin) that it took a full fifty years for the Newtonian scheme of the universe to be widely accepted and taught in schools and universities. Even at Cambridge itself, six years after the *Principia* was published, the professors were still teaching the Cartesian system, based on the French mathematician Descartes's incorrect theories of the solar system developed in the early 1600s. When the second edition of the *Principia* was published in 1713, which included new material on the motions of the moon, comets, and other heavenly bodies and on the resistance of fluids, Newton's associate Roger Cotes still was not able to convince critics that Descartes must be abandoned in favor of Newtonian physics.

Newton Loses Interest in Science and Broadens His Life

The Prophesies of Daniel and Paranoid Delusions

For several years after the publication of the *Principia*, Newton displayed very little interest in scientific matters. The long and tedious process of getting the work published had left him exhausted and indifferent to science, similar to an earlier period when Newton was in his mid-thirties and wrote to a friend that his interest in science was "worn out." In 1689, he was elected to Parliament and served for one year, a position that brought him to London often, where he became friends with the great philosopher John Locke and many men of science who would later play an influential role in Newton's remaining years. During these post-*Principia* years, he wrote extensively about the Scriptures, including correspondence with John Locke on the prophesies of Daniel.

After failing to be appointed to several high government posts that he sought, Newton finally was named warden of the mint in March 1696. In 1699 he was promoted to master of the mint, where he succeeded in resurrecting the seriously debased coinage of the British empire and where he received much recognition as an innovative administrator. He retained that position until his death in 1727. At about the time he assumed his position at the mint in 1699, Newton exhibited a renewed interest in science. In 1703, he was elected president of the Royal Society, the most prestigious science organization in Europe at the time, a position he held until his death. He was the first distinguished man of science to hold that office since the astronomer Christopher Wren (1632–1723) twenty years earlier.

In 1704, Newton published his greatest work on optics, called *Opticks: or a Treatise on the Reflexions, Refractions, Inflexions and Colours of Light*. It included two mathematical treatises that he had written years earlier to support his optics theories. In much the same way that the release of the *Principia* was long delayed, all the notes for the *Opticks* were prepared even before Newton came to London in 1689, and as Newton stated in the preface, mostly written in 1675. Primarily because of the publication of the *Opticks*, Newton was even more exalted throughout the scientific world. He was knighted in 1705, an honor that had never before been conferred for achievements in science.

Unlike most other great men of science, Newton continued to have long periods when he was not interested in physics, during which he devoted all his time and writing to chemistry, alchemy, his duties at the mint, and religious matters. Much of his religious writing was collected and published after his death, including *An Historical Account of Two Notable Corruptions of the Scriptures*, *The Chronology of Ancient Kingdoms*, and *Observations Upon the Prophecies of Daniel and the Apocalypse of St. John*. On the subject of chemistry and alchemy, Newton left behind a large library of books. In addition, he engaged in extensive correspondence with his friend Robert Boyle, the famous English chemist, concerning matters of chemical combinations and experiments. In keeping with Newton's secretive personality, he never revealed the purpose or results of his own chemical experiments.

Obviously, even great men are not perfect. Isaac Newton was easily irritated, overly sensitive to the slightest criticism of his work, intent on gaining revenge on his enemies (both real and imagined), and impatient with people of questionable talent or low motivation. Between 1689 and 1692, after he'd served in Parliament and when his interest in science was in one of its low ebbs, he repeatedly accused his close associates and friends

John Locke and Charles Montagu of deceiving him and attempting to harm his reputation. He recovered from these paranoid delusions by 1693. He seemed to relish his public disputes with other scientists, such as the British astronomer John Flamsteed, who challenged Newton's work in optics and astronomy, and the German mathematician Gottfried Wilhelm Leibniz, whom Newton accused of plagiarizing his own work in optics and the development of the differential calculus. In the first year of his presidency of the Royal Society, Newton wrote "our Society decays and produces nothing remarkable . . . 'tis governed by persons that either value nothing but their own interests, or understand little but vegetables" (referring to Sir Hans Sloane, the secretary of the Royal Society). These disputes and others flamed throughout Newton's life and were either the cause or the effect of his embitterment.

Although this book doesn't attempt to rank the seven greatest scientific discoveries in history or their discoverers in any particular order, it is appropriate that Isaac Newton comes first. Newton was not only first chronologically (with five of the remaining six great discoveries taking place in the twentieth century), but much of what was discovered after him would not have been possible without his work. The impact of Newton's discoveries continues in the twenty-first century in many fields of science. The wave theory of light uses Newton's laws of motion, as does the kinetic theory of heat. Indeed, three centuries after Newton published *Opticks*, scientists have now developed "metamaterials" that work by steering light and other electromagnetic waves. This produces far superior resolution than ordinary lenses, which have an "index of refraction" greater than one. Index of refraction of a material (like a glass lens) is the ratio of the speed of light in a vacuum to the speed of light in that material, with the ratio typically being greater than one. In 2012, scientists developed a device out of silicon dioxide and silver with an index of refraction of zero, the first time for visible light. The practical uses for such a groundbreaking device remain to be discovered by modern physicists. Newtonian theory also was important in the development of our understanding of electricity and magnetism and in the discoveries of electrodynamics and optics by Faraday and Maxwell. Newton's physics guided science for over 200 years, until the early twentieth century when Einstein demonstrated that physics needed to grow beyond the Newtonian framework. Because Isaac Newton supplied the basic principles on which much of science has been built and led the transformation that brought modern science into existence, he will remain the supreme symbol of the power of the human mind to apply reason and the scientific method to understand how the physical universe operates.

The Structure of the Atom

In Focus

Two hundred fifty million hydrogen atoms, packed together side by side, would be about one inch long. If an atom were blown up to the size of the Superdome in New Orleans, the nucleus (made of protons and neutrons) would be the size of a pearl in the center of the structure, with the electrons orbiting at the speed of thousands of miles per second at the outer perimeter of the building.

The nucleus is extremely heavy, accounting for 99.95 percent of the total weight of the atom, with each proton having a mass about 1,836 times greater than each electron. That is, these electrical charges *weigh* something! However, the "parts" or subatomic particles that make up each atom of every element (that is, hydrogen, oxygen, gold, silver, aluminum, uranium, and all the other chemical elements) are colorless, tasteless, odorless, and without texture. The subatomic particles are not hard, soft, dull, or shiny even though the ninety-eight naturally occurring elements make up everything that exists on Earth—oceans, rocks, air, and all plants and animals, including us. The *firmness* and other characteristics of these everyday objects made of this "stuff" results from the interrelationship of electrical forces of and between subatomic particles.

Matter in its most fundamental form consists solely of these electrical charges. This substance from "nothing" that results when electrical charges combine to form atoms and when atoms combine to form elements and molecules would qualify as the greatest illusion there is—except it is reality.

Part Two discusses how scientists discovered this staggering vision of the complexity of the physical world, identifies those primarily responsible for that dramatic discovery, and explores the incredible story of how we came to unearth the key that released the enormous energy locked within the atomic nucleus.

Chapter Four
GOOD CHEMISTRY

First we must inquire whether the elements [earth, air, fire and water] are eternal or subject to generation and destruction. . . . It is impossible that the elements should be generated from some kind of body. That would involve a body distinct from the elements and prior to them.

—Aristotle, *On the Heavens* (c. 330 BCE)

In *The Ascent of Man*, Jacob Bronowski wrote, "The genius of people like Newton and Einstein is that they ask simple, innocent questions which turn out to have answers with enormous effects on society and individuals." Asking such questions is an ancient tradition begun in Greece, which was first inhabited 8,000 years ago and was the first enduring civilization in Europe. Aristotle coined the term *physics*, from the Greek word *physis* ("nature"), to designate the study of nature. Though most of the teachings and theories of Socrates, Plato, and Aristotle were proved inaccurate and were abandoned during the Renaissance, some of their philosophies persisted for nearly 2,000 years. Modern science can trace its roots back to ancient Greece not because these or other ancient Greek philosophers made any scientific discoveries, but because they believed the physical world and universe could be understood through rational thought. This reliance on logic and reason ultimately grew into the scientific method of modern science that was applied so well by Newton. Beginning in the same era as Newton's great work, other scientists were attempting to understand the composition of matter at its most fundamental level. This exploration would take centuries to reach fruition but actually had its origin in the "physis" of ancient Greece.

Modern Atomic Theory Traces Its Roots to Ancient Greece

Sound and Souls Made of Atoms

In about 500 BCE, the Greek philosopher Anaxagoras inquired into what matter is made of and imagined cutting in half any bit of matter—say, a ball of silver—then halving it again and again and again. After contemplating this, he proposed his theory that a person could continue halving any piece of matter without ever reaching an end. In other words, all matter is infinitely divisible.

On the northern shore of the Aegean Sea, there was the Greek town Abdera (now Avdhira), founded by refugees from the Persian invasion of Lydia in the year 540 BCE. A man named Leucippus settled there in 478 BCE and had a theory about matter that was fundamentally different from Anaxagoras's conclusions and from anything else espoused prior to that time. Only fragments of his writings remain, so we have to rely on the writings of his pupil Democritus, who popularized Leucippus's thoughts, and on Aristotle's characterization of Leucippus's philosophies, which are often referred to by Aristotle in his voluminous writings. Leucippus's theory stated that two things and only two things exist in the universe: *atoms and the void*. Leucippus believed that everything we see is made of atoms (in Greek *a-tomas*, meaning "not cuttable") that move through the void of empty space. He conjured up these hypothetical bits of matter and said they're too small to be seen; they're solid, so solid that they cannot be divided in any way (stable, unchangeable, and indestructible); they occur in various geometric forms (which explains their capacity to combine to form all the things in the world); and *they are perpetually in motion*. They form the material that we see, through their collisions and regroupings. He couldn't imagine that this magnificent universe is made from formless nothingness with no basic identity. Otherwise, how could material things retain their individuality and how could living things pass along their complex images to their offspring without some type of indestructible pieces of matter?

Leucippus's brilliant and prescient atomic theory was later used by other philosophers to explain a wide variety of phenomena. For example, sound was supposedly generated by atoms of sound hitting atoms of air, which then struck the ear. Even the human soul was supposedly composed of atoms. As with almost all the enduring principles handed down through ancient Greece, Leucippus's atomic theory and others' application of it was totally based on speculation. It was philosophy not science. But to Leucippus's credit, he proposed a concept that resurfaced 2,400 years later in the modern understanding of the structure of the atom.

Others in ancient Greece pondered the nature of matter before and after Leucippus and Anaxagoras. Empedocles (c. 495–435 BCE) was the first to propose that earth, air, fire, and water were the elemental substances, while a host of other Greek philosophers attempted to understand and explain our physical surroundings. Thales of Miletus (sixth century BCE) proposed that water was the basis of all matter. Anaximander of Miletus (c. 610–545 BCE) spoke vaguely of a "primary substance" that had the capacity of all types of matter and their properties (hot, cold, wet, dry, etc.). Anaximines of Miletus (c. 585–524 BCE) proposed that air is the primary substance and transforms into other materials. Plato and Aristotle adopted Empedocles's four elements, and added their personal spin on how matter and the physical universe are put together. Though people adhered to the belief in the four elements through the Middle Ages, Leucippus's atomic theory would ultimately succeed that belief.

Procedures and Apparatus of Medieval Alchemy Lay Foundation for Modern Chemistry

Yet Earth, Air, Fire, and Water Persist

Beginning thousands of years ago, people valued gold as a rare and beautiful substance, and also recognized that gold had a unique ability to resist decay and corrosion. Because there was no known acid or other substance that could damage gold, it was thought to have a quality of permanence that could be transmitted to humans. Therefore, every medicine for fighting aging contained gold as an essential ingredient, and doctors urged people to drink from gold cups to prolong life.

This universal desire for gold spawned alchemy as a formal discipline in the first century CE. It first arose among Greek scholars, then spread to eastern Mediterranean countries, and finally reached Spain and Italy in the twelfth century. Though the attempt to produce gold from other substances was the original and central purpose of alchemy, a number of physician-alchemists in Europe in the Middle Ages attempted to produce medicines that were not dependent on gold or related to it. In an effort to produce medicinal essences and spirits from raw materials, such as herbs, this group of alchemists improved methods of separating elements by distillation. For example, as early as the thirteenth century, Thaddeus of Florence identified the medical benefits of alcohol distillates taken internally and applied topically. Paracelsus (1493–1541), the German-Swiss physician and alchemist, was the first person to unite medicine with chemistry through his use of remedies that contained mercury, sulfur, iron, and copper sulfate. This led to the use of steam distillation and improved equipment.

The development of apparatus and the extensive efforts to break down or distill substances laid the foundation for modern chemistry, but as true science began to evolve during the Renaissance, the study of alchemy *impeded* the birth of modern chemistry. Even as Francis Bacon's teachings on the scientific method led people away from Greek philosophy toward reliance on empirical evidence (that is, what can actually be observed and/or measured), the four elements (earth, air, fire, and water) lived on, and it was not yet recognized that these four substances actually are composed of a *combination* of the basic elements of matter.

Greek Philosophy Gives Way to the Scientific Method as Elements Are Discovered

Phlogiston and Guillotines

Chemistry—the laws that govern the behavior of the elements—finally broke free from alchemy in 1661 when the Irish chemist Robert Boyle (1627–1691) published *The Sceptical Chymist*, in which he demonstrated that there was no basis for regarding earth, air, fire, and water as elemental. Boyle, one of the founding members of the Royal Society of London in 1660, laid the foundation for qualitative analysis using flame colors, spot tests, precipitates, and other analytical tools in an effort to figure out the basic elements making up minerals and other materials. In 1662, he determined that the volume of a gas is inversely proportional to the pressure exerted on that gas. So, for example, doubling the pressure on a particular volume of gaseous substance causes it to become *half* that volume, and reducing the pressure to half of the original pressure causes it to *double* in volume. This became known as Boyle's Law and led to his thesis that gases must be made up of tiny "corpuscles" with a large amount of empty space between them, which is the reason they can be squeezed into smaller volumes. Isaac Newton concurred with Boyle's corpuscle explanation when he referred to atoms as "the small particles of bodies with certain powers, virtues, or forces" by which they could act on each other. With this concept, Boyle declared that Leucippus must have been correct—that these corpuscles are comparable to Leucippus's atoms. Having resurrected Leucippus's version of indivisible particles, Boyle became one of the first significant figures in science to turn his back on Anaxagoras's and Aristotle's teachings about infinitely divisible matter and to finally shatter the centuries-old persistent belief that earth, air, fire, and water are the fundamental materials of the universe.

Joseph Black (1728–1799), born in France and educated in Scotland, began experimenting on the constituents of air in the 1750s, which led to his discovery of carbon dioxide and nitrogen. Beginning with Black, scores of other European chemists were attracted by the curious and unique properties of gases and began to focus on their makeup. These studies led to the realization that there is a group of gaseous substances unique from one another that *combine* to make up "air." Therefore, the concept of air as *one element* gave way to "gas" being considered a *state of matter*, along with liquid and solid states, today's proven concept. This thinking rapidly led to the further realization that a combination of several fundamental substances make up all material on Earth.

In 1729, Georg Ernst Stahl, physician to the king of Prussia, had "invented" the concept of phlogiston, a mysterious, invisible, colorless, odorless, and tasteless substance that supposedly had "negative weight." Stahl proposed phlogiston to explain why certain substances became heavier when burned and as an excuse to cover up the increasing failings of Aristotelian theories to which many people still clung. Supported by a number of leading chemists, the phlogiston theory persisted for decades. For example, the British chemist Joseph Priestley (1733–1804) was the first to isolate oxygen (in 1774) and describe its role in combustion and respiration, but he called it "dephlogisticated air" and didn't understand the full significance of the discovery. This was left to Lavoisier.

Antoine Laurent Lavoisier (1743–1794), the son of a wealthy Parisian lawyer, pursued the family tradition and received his license to practice law in 1764. But within two years he was drawn back to his desire to learn more about science, an interest first experienced during his earlier education in math, astronomy, chemistry, and botany. By 1772 he'd disproven several of the ancient Greek principles about earth, air, fire, and water, and developed a reputation for exact quantitative procedures and brilliant experiments. Lavoisier possessed the insight and vision necessary to get over the phlogiston hump through his careful measurement of the weight of substances to determine the effect of heating them and his development of the principle that *a substance could be considered elemental only if it failed to break down into simpler substances when treated chemically.* He expanded the list of known elements to thirty-three, although some were erroneous. From 1776 to 1782, Lavoisier conducted experiments in which he isolated oxygen in air and furthered Priestley's work on oxygen's role in combustion and respiration. The term *oxygine* (Greek for "acid former") was used for the first time in a memoir by Lavoisier, dated September 5, 1777. In a 1783 paper (succinctly titled "On the Nature of Water and on Experiments That Appear to Prove That This Substance Is Not Properly Speaking an Element, but Can Be Decomposed and Recombined"), Lavoisier reported to the French Academy of Sciences that water was the product of combining hydrogen and oxygen. In a subsequent paper delivered to the academy, Lavoisier presented a logical analysis about the substance that we now call "oxygen." Through Lavoisier's sensitive balance instruments, keen insight, and inductive reasoning, he completed Boyle's efforts to vanquish the ancient Greek concept of earth, air, fire, and water once and for all, and he put the tenacious phlogiston theory to rest. For this and other work, Lavoisier is now considered the father of modern chemistry.

Lavoisier had been active in political affairs his entire adult life, and devoted much of his career to public service, including positions in the French government from 1768 to 1790 in the areas of economics, agriculture, education, and social welfare. In the aftermath of the French Revolution of 1789, despite his many contributions as a reformer and political liberal and despite his participation in the revolution, he came under attack because of his status as a wealthy member of the French aristocracy, primarily because of a position he'd held in 1768 in the Ferme Générale, the country's tax collecting agency. When the Reign of Terror commenced in 1793, resulting in the suppression of the French Academy of Sciences and other learned societies, Lavoisier was arrested. On May 8, 1794, after a one-day trial, the prospect of Lavoisier's further contributions to science and rational thought were prematurely halted at age fifty-one at Place de la Concorde in Paris, as his great mind fell into the blood-soaked basket at the foot of the guillotine along with twenty-seven other former members of the Ferme Générale.

Dalton Develops the First Modern Theory of Atoms

Ultimate Particles

But what controls the properties of the elements that make up all matter? Remarkably, the son of a poor Quaker weaver in rural England answered that question when he established the idea of chemical atomism. In October 1803, John Dalton (1766–1844), professor of mathematics and physical sciences at New College, Manchester, delivered a paper to the Manchester Literary and Philosophical Society concerning the mixture and solubility of different gases: "The circumstance depends upon the weight and number of the ultimate particles of several gases. . . . An enquiry into the relative weights of the ultimate particles of bodies is a subject, as far as I know, entirely new. I have lately been prosecuting the enquiry with remarkable success." One month before this speech he'd prepared a table of comparable atomic weights, based largely on calculations made by Lavoisier and others.

By the first decade of the 1800s, we had come full circle from Leucippus in Greece in 478 BCE to John Dalton in England. Now there was a theory—more refined than Boyle's corpuscle theory—derived from two centuries of chemical experiments in which Dalton reached these conclusions:

- The elements are made up of tiny indivisible particles, which Dalton called "atoms" in honor of Leucippus and Democritus.

FIGURE 4-1

Dalton's 1803 Table of the Relative Weights
of the Ultimate Particles

Hydrogen	1.0	Nitrous oxide	13.7
Azot	4.2	Sulphur	14.4
Carbone	4.3	Nitric acid	15.2
Ammonia	5.2	Sulphuretted hydrogen	15.4
Oxygen	5.5	Carbonic acid	15.3
Water	6.5	Alcohol	15.3
Phosphorous	7.2	Sulphureous acid	19.9
Phosphuretted hydrogen	8.2	Sulphuric acid	25.4
Nitrous gas	9.3	Carburetted hydrogen	6.3
Ether	9.6	Olefiant gas	5.3
Gaseous oxide of carbone	9.8		

- The atoms of each element are all alike but differ from the atoms of every other element.

- Chemical combination occurs when the atoms of two or more elements form a "firm union."

It hadn't been necessary for Kepler or Newton to understand the atom to figure out the laws of gravity or motion. But people searching for the ultimate structure of matter in the early 1800s began to realize that they were on the path to understanding and unlocking this secret. The Swedish physician and chemist Jöns Jacob Berzelius (1779–1848) began his research in analytical chemistry soon after he became aware of Dalton's atomic theory and was one of the first scientists to recognize its great significance. Beginning in 1810, Berzelius made major contributions in determining accurate atomic weights and created the standard system of chemical symbols and chemical proportions still in use today.

Over the next few decades, Dalton's atomic theory was gradually accepted by other scientists and a greater understanding of "firm unions" that result in molecules was realized through the work of the Italian physicist Amedeo Avogadro (1776–1856) and the Italian chemist Stanislao Cannizzaro (1826–1910), whose combined theories provided a means to determine molecular weight and size. Introduced in 1811, Avagadro's two-part hypothesis held that (1) *the ultimate particles are not necessarily atoms but may be groups of atoms joined to form molecules* and (2) *equal volumes of gases contain equal numbers of molecules*. Despite its accuracy and usefulness, Avagadro's hypothesis was not consistent with the accepted principles of that time. Therefore, it went largely unnoticed until Cannizzaro developed

FIGURE 4-2
Discovery of the Elements

Before 1700	1700–1799	1800–1849	1850–1899
Antimony	Beryllium	Aluminum	Actinium
Arsenic	Bismuth	Barium	Argon
Carbon	Chlorine	Boron	Cesium
Copper	Chromium	Bromine	Dysprosium
Gold	Cobalt	Cadmium	Gadolinium
Iron	Fluorine	Calcium	Gallium
Lead	Hydrogen	Cerium	Germanium
Mercury	Manganese	Erbium	Helium
Phosphorus	Molybdenum	Iodine	Holmium
Silver	Nickel	Lanthanum	Indium
Sulfur	Nitrogen	Iridium	Krypton
Tin	Oxygen	Lithium	Neodymium
	Platinum	Magnesium	Neon
	Strontium	Niobium	Polonium
	Tellurium	Osmium	Praseodymium
	Titanium	Palladium	Radium
	Tungsten	Potassium	Rhodium
	Uranium	Rubidium	Ruthenium
	Yttrium	Selenium	Samarium
	Zinc	Silicon	Scandium
	Zirconium	Sodium	Thallium
		Tantalum	Thulium
		Thorium	Xenon
		Vanadium	Ytterbium

additional evidence to support it and presented the hypothesis to a congress of chemists in Germany in 1858. After this, chemistry rapidly matured into a true science and the elements were organized into a new conceptual framework and system.

Mendeleev Constructs a Conceptual Framework for the Elements

Gaps in the Table

By 1869 sixty-one elements were known. Dmitri Mendeleev (1834–1907), born in Siberia, Russia, the youngest of seventeen children of a schoolteacher, prepared a set of cards in which the properties (atomic weight, specific gravity, volume, valence, specific heat, etc.) of each of these elements were tabulated on a separate card. By arranging and rearranging the cards, Mendeleev figured out that such properties recurred "periodically" when the cards were arranged on the basis of increasing atomic weight. He then pre-

FIGURE 4-3
Replica of Mendeleev's Vertical Table of 1869

			K = 39	Rb = 85	Cs = 133	—	—
			Ca = 40	5r = 87	Ba= 137	—	—
			—	?Yt = 88?	?Di = 138?	Er = 178?	—
			Ti = 48?	Zr = 90	Ce = 140?	?La =180?	Th = 231
			V = 51	Nb = 94	—	Ta = 182	—
			Cr = 52	Mo = 96	—	W = 184	U = 240
			Mn = 55	—	—	—	—
Typische		—	Fe = 56	Ru = 140	—	Os = 195?	
Elements		—	Co = 58	Rb = 104	—	Ir = 197	
			Ni = 59	Pd = 105	—	Pt = 198?	—
H = 1	Li = 7	Na = 23	Cu = 63	Ag = 108	—	Au = 199?	—
	Be = 9,4	Mg = 24	Zn = 66	Cd = 112	—	Hg = 200	—
	B = 11	Al = 27,3	—	In = 113	—	T1 = 204	—
	C = 13	Si = 28	—	Sn = 118	—	Pb = 207	—
	N = 14	P = 31	AS = 75	Sb = 122	—	Bi = 208	—
	O = 16	S = 32	Se = 78	Te = 125?	—	—	—
	F = 19	Cl = 35,5	Br = 80	1 = 127	—	—	—

pared a table to reflect this principle, which was published in 1869, as shown in Figure 4-3.

Mendeleev developed principles that were applied in the table, including listing according to increasing weight, and placing the symbol for hydrogen away from the elements that followed hydrogen in weight. Because he had the insight to see that many elements had not yet been discovered, he left open spaces in the periodic table—he imagined the missing pieces, based on the "color and shape" of those that were there, like in a jigsaw puzzle. For example, he predicted that an unknown element would be found for the space following calcium and that it would be closely related to boron. This prediction was proven correct in 1879 when the Swedish chemist Lars Fredrik Nilson (1840–1899) discovered scandium. Mendeleev's table devel-

FIGURE 4-4
The Modern Periodic Table

Group → Period ↓	1	2	3	4	5	6	7	8	9	10	11	12	13	14	15	16	17	18
1	1 H																	2 He
2	3 Li	4 Be											5 B	6 C	7 N	8 O	9 F	10 Ne
3	11 Ne	12 Mg											13 Al	14 Si	15 P	16 S	17 Cl	18 At
4	19 K	20 Ca	21 Sc	22 Ti	23 V	24 Cr	25 Mn	26 Fe	27 Co	28 Ni	29 Cu	30 Zn	31 Ga	32 Ge	33 As	34 Se	35 Br	36 Kr
5	37 Rb	38 Sr	39 Y	40 Zr	41 Nb	42 Mo	43 Tc	44 Ru	45 Rh	46 Pd	47 Ag	48 Cd	49 In	50 Sn	51 Sb	52 Te	53 I	54 Xe
6	55 Cs	56 Ba		72 Hf	73 Ta	74 W	75 Re	76 Os	77 Ir	78 Pt	79 Au	80 Hg	81 Tl	82 Pb	83 Bi	84 Po	85 At	86 Rn
7	87 Fr	88 Ra		104 Rf	105 Db	106 Sg	107 Bh	108 Hs	109 Mt	110 Ds	111 Rg	112 Cn	113 Uut	114 Fl	115 Uup	116 Lv	117 Uus	118 Uuo

Lanthanides	57 La	58 Ce	59 Pr	60 Nd	61 Pm	62 Sm	63 Eu	64 Gd	65 Tb	66 Dy	67 Ho	68 Er	69 Tm	70 Yb	71 Lu
Actinides	89 Ac	90 Th	91 Pa	92 U	93 Np	94 Pu	95 Am	96 Cm	97 Bk	98 Cf	99 Es	100 Fm	101 Md	102 No	103 Lr

This is an eighteen-column periodic table layout, referred to as the common, or standard, form. It is also sometimes referred to as the long form. The wide periodic table incorporates the lanthanides and the actinides, rather than separating them from the main body of the table. The extended periodic table adds the eighth and ninth periods, including superactinides.

oped into the modern periodic table, one of the most important tools in chemistry. The vertical columns of the modern periodic table are called *groups* and the horizontal rows are called *periods*. The atomic number of an element is the number of protons in the nucleus of the atoms of that element. Neutrons are ignored when determining atomic number and weight.

Not only does the modern periodic table succinctly organize all of the elements but it clearly illustrates that they form "families" in rational groups or trends, based on their salient characteristics (atomic weight, diameters, density, and energy).

Figure 4-5 lists all of the ninety-eight naturally occurring elements in order of atomic number (1 hydrogen through 98 californium). Until about 2003, the generally accepted number of naturally occurring elements was ninety-two. There is still some disagreement as to whether there are ninety-four or ninety-eight natural elements because several break down (decay) so rapidly they are difficult to detect. But this number does not change the concept of an element. Thus the table also gives the atomic weight, which is based on a scale that uses a specific type of carbon atom (carbon-12) as the standard against which all other elements' weights are measured. Carbon-12 is an *isotope* of the carbon atom, meaning it is one phase of carbon among others of nearly identical chemical behavior but with different atomic masses. Atoms of the same element that have different numbers of neutrons are known as isotopes of that element. The majority of elements are mixtures of several isotopes—that is, the ratio of neutrons to protons varies slightly. The atomic weights of elements with no "stable" (that is, nondecaying) isotopes are listed in brackets in Figure 4-5 as the mass of their most stable isotope. A number of elemental isotopes are central to the essential technique known as radiometric dating. In chapters to follow, we will learn that ancient rocks, bones, and fossils have given us a window to the earliest pages of Earth's and humankind's history. The age of these ancient remnants is determined by comparing the observed abundance of a naturally occurring radioactive isotope and its decay products, using known decay rates. For example, carbon-14 is a radioactive isotope of carbon-12, with a half-life of 5,730 years, which is very short compared with other isotopes. This means that half of the molecules of carbon-14 will be lost every 5,730 years. Thus, the carbon-14 dating limit works best for organic materials less than around 60,000 years old. Carbon-14 is continuously created through collisions of

FIGURE 4-5

List of Naturally Occurring Elements

Atomic No.	Name	Symbol	Weight	Atomic No.	Name	Symbol	Weight
1	Hydrogen	H	1.00794(7)	37	Rubidium	Rb	854678(3)
2	Helium	He	4.002602(2)	38	Strontium	Sr	87.62(1)
3	Lithium	Li	6.941(2)	39	Yttrium	Y	88.90585(2)
4	Beryllium	Be	9.012182(3)	40	Zirconium	Zr	91.224(2)
5	Boron	B	10.811(5)	41	Niobium	Nb	92.90638(2)
6	Carbon	C	12.011(1)	42	Molybdenum	Mo	95.94(1)
7	Nitrogen	N	14.00674(7)	43	Technetium	Tc	[98]
8	Oxygen	O	15.9994(3)	44	Ruthenium	Ru	101.07(2)
9	Fluorine	F	18.9984032(9)	45	Rhodium	Rh	102.90550(3)
10	Neon	Ne	20.1797(6)	46	Palladium	Pd	106.42(1)
11	Sodium	Na	22.989768(6)	47	Silver	Ag	107.8682(2)
12	Magnesium	Mg	24.3050(6)	48	Cadmium	Cd	112.411(8)
13	Aluminum	Al	26.981539(5)	49	Indium	In	114.818(3)
14	Silicon	Si	28.0855(3)	50	Tin	Sn	118.710(7)
15	Phosphorous	P	30.973762(4)	51	Antimony	Sb	121.760(1)
16	Sulfur	S	32.066(6)	52	Tellurium	Te	127.60(3)
17	Chlorine	Cl	35.4527(9)	53	Iodine	I	126.90447(3)
18	Argon	Ar	39.948(1)	54	Xenon	Xe	131.29(2)
19	Potassium	K	39.0983(1)	55	Cesium	Cs	132.90543(5)
20	Calcium	Ca	40.078(4)	56	Barium	Ba	137.327(7)
21	Scandium	Sc	44.955910(9)	57	Lanthanum	La	138.9055(2)
22	Titanium	Ti	47.867(1)	58	Cerium	Ce	140.115(4)
23	Vanadium	V	50.9415(1)	59	Praseodymium	Pr	140.90765(3)
24	Chromium	Cr	51.9961(6)	60	Neodymium	Nd	144.24(3)
25	Manganese	Mn	54.93805(1)	61	Promethium	Pm	[145]
26	Iron	Fe	55.845(2)	62	Samarium	Sm	150.36(3)
27	Cobalt	Co	58.93320(1)	63	Europium	Eu	151.965(9)
28	Nickel	Ni	58.6934(2)	64	Gadolinium	Gd	157.25(3)
29	Copper	Cu	63.546(3)	65	Terbium	Tb	158.92534(3)
30	Zinc	Zn	65.39(2)	66	Dysprosium	Dy	162.50(3)
31	Gallium	Ga	69.723(1)	67	Holmium	Ho	164.93032(3)
32	Germanium	Ge	72.61(2)	68	Erbium	Er	167.26(3)
33	Arsenic	As	74.92159(2)	69	Thulium	Tm	168.9342(3)
34	Selenium	Se	78.96(3)	70	Ytterbium	Yb	173.04(3)
35	Bromine	Br	79.904(1)	71	Lutetium	Lu	174.967(1)
36	Krypton	Kr	83.80(1)	72	Hafnium	Hf	178.49(2)

Atomic No.	Name	Symbol	Weight	Atomic No.	Name	Symbol	Weight
73	Tantalum	Ta	180.9479(1)	86	Radon	Rn	[222]
74	Tungsten	W	183.84(1)	87	Francium	Fr	[223]
75	Rhenium	Re	186.207(1)	88	Radium	Ra	[226]
76	Osmium	Os	190.23(3)	89	Actinium	Ac	[227]
77	Iridium	Ir	192.217(3)	90	Thorium	Th	232.0381(1)
78	Platinum	Pt	195.08(3)	91	Protactinium	Pa	231.03588(2)
79	Gold	Au	196.96654(3)	92	Uranium	U	238.0289(1)
80	Mercury	Hg	200.59(2)	93	Neptunium	Np	237
81	Thallium	Tl	204.3833(2)	94	Plutonium	Pu	244
82	Lead	Pb	207.2(1)	95	Americium	Am	243
83	Bismuth	Bi	208.98037(3)	96	Curium	Cm	247
84	Polonium	Po	[209]	97	Berkelium	Bk	247
85	Astatine	At	[210]	98	Californium	Cf	251

neutrons generated by cosmic rays and thus remains at a near-constant level on Earth. A living plant or animal acquires carbon throughout its lifetime through photosynthesis and from consumption of plants and other animals. When an organism dies, it ceases to take in new carbon-14, and the existing isotope decays with the 5,730-year half-life. The amount of carbon-14 that remains when an organism is examined provides an indication of the time elapsed since its death. To establish the age of rocks and fossils that may be millions of years old, scientists use radiometric dating by analyzing isotopes of uranium-lead or potassium-argon that have half-lives of millions to billions of years. For example, one isotope of uranium takes 700 million years to decay to an isotope of lead. One isotope of potassium has a half-life of 1.3 billion years and can be used to date the oldest existing rocks and fossils.

There are twenty manmade elements in the periodic table. These have not been demonstrated to occur naturally in the universe but have been synthesized as a product of manmade nuclear reactions and given official names by the International Union of Pure and Applied Chemistry (IUPAC). The first of these produced was einsteinium (Es; atomic weight: 99), in 1952, followed by fermium (Fm; 100), that same year. The element mendelevium (Md; 101) was next, followed by nobelium (No; 102), lawrencium (Lr; 103), rutherfordium (Rf; 104), dubnium (Db; 105), seaborgium (Sg; 106), bohrium (Bh; 107), hassium (Hs; 108), meitnerium (Mt; 109), darmstadtium (Ds; 110), roentgenium (Rg; 111), and copernicium (Cn; 112). Numbers 110, 111, and 112 were finally named by the IUPAC in 2011, and numbers 114 (flerovium; Fl) and 116 (livermorium; Lv) were named in 2012. As of this writing, the following elements have temporary labels: ununtrium (Uut; 113), ununpentium (Uup; 115), ununseptium (Uus; 117), and ununoctium (Uuo; 118); the latter is the element with the highest atomic number and mass produced so far (see Figure 4-4).

The year 2011 was declared the International Year of Chemistry to celebrate the field's volumes of contributions. It is interesting to note that as chemists developed and named the various elements, the luminary Glen Seaborg (see element 106) proposed the names pandemonium and delirium for elements 95 and 96 (respectively, americium and curium), reportedly because of the great difficulties and stresses involved in pursuing this path of chemistry. The history of chemistry is fascinating, and the importance of this field is enormous beyond words, but scientists soon realized that to understand fully the physical basis for the laws of chemistry they would have to determine the actual behavior of the atoms that make up every one of the known elements. This was considered impossible until late in the nineteenth century. Also, for most physicists in the late 1800s, Dalton's description of the atom fell short, mostly because it didn't seem to fit into the Newtonian system of physical principles and didn't explain the electrical behavior of matter. One line of inquiry regarding such electrical behavior was physicists' efforts to better understand the behavior of gases when combined with electricity. Numerous experiments on these phenomena were conducted by physicists and chemists throughout the nineteenth century. In 1879, the British chemist and physicist William Crookes (1832–1919) reported the results of his extensive research on electrical discharges through gas. In his experiments, he used a vacuum tube of his own design (now known as a Crookes tube), which included a cathode (that is, a negative electrode) and an anode (a positive electrode), thus allowing a charge to pass through the tube. Crookes noted that rays of unknown origin and composition were coming from the tube when electricity passed through the gas in the tube. He was able to determine several characteristics of these rays, in that they travel in straight lines and produce phosphorescence and heat in certain materials. Over the next fifteen years, a number of other chemists and physicists experimented with so-called cathode rays and Crookes tubes in an effort to understand the composition and characteristics of the mysterious emanations.

As scientists approached the twentieth century there was much greater motivation for discovering and understanding what matter is made of than there had been in 478 BCE when Leucippus first put forth his crude atomic theory. When it became apparent that all atoms shared certain characteristics and that the atom might even be composed of smaller components, physicists and chemists seized the baton from their predecessors who were active in the later decades of the nineteenth century. In the following years they dramatically unraveled the internal structure of the atom. As shown in the next chapter, the effort to understand the nature of cathode rays created a direct link to the discovery of atomic structure. Then, through a degree of serendipity, indirect observation of the atom's behavior, and much painstaking research, the "pieces" of the atom were discovered, enabling scientists to figure out their peculiar, unique, and once secret laws.

Chapter Five
QUANTUM LEAP

The atom revealed its secrets only gradually, like a cleverly written play, allowing sparks of knowledge briefly to illuminate corners of the story but jumbling scenes and inferences so that full realization came only at the end, after many members of the audience had given up hope of understanding.
—Robert P. Crease and Charles C. Mann, *The Second Creation* (1986)

From a distance, beach sand appears to be one piece of homogeneous material. Simply because human eyesight is incapable of discerning each grain of sand from a distance away, no one would claim that the beach is a single entity or mass without constituent pieces. Likewise, despite our inability to see atoms we have no basis for rejecting any particular configuration of them in the absence of proof. Seeing is not necessarily believing. If common sense suggests that the atom is made out of matter, then what "kind" of matter is it? Is that tiny piece of "matter" infinitely divisible, as Anaxagoras believed? Why? Into what? Is there a *smallest* piece of matter?

Röntgen's Accidental Discovery of X-Rays Begins the Search for Subatomic Particles

Rocks Beaming Out Rays . . . Plum Pudding

If a date could be set for the birth of nuclear physics, it would be November 8, 1895. On that day, Wilhelm Conrad Röntgen (1845–1923), a physics professor at Wurzburg, Germany, was studying the emission of ultraviolet light from a cathode ray tube by directing the light at crystals of barium platinocyanide, a chemical compound that combines the element barium (atomic number 56) with the elements platinum (78) and cyanide, which is made up of a carbon atom (6) triple-bonded to nitrogen (7). The chemists used this compound because it fluoresces under ultraviolet light. That is, the crystals seem to absorb the ultraviolet light and then return it by giving off their own light.

In the course of his experiments, Röntgen noticed that the rays emanating from the tube were penetrating *through* various substances and objects across the room, several feet away from the tube, and causing the crystals to fluoresce. After investigating this for the next several weeks and eliminating all other possible causes, he directed the rays at sensitive photographic plates and was able to register the image of his wife's bones in her hand, because the rays (which he called "X," the scientific symbol for the unknown) passed through the softer skin and muscle in her hand more readily than they passed through the bones, thus producing essentially a shadow of the bones in her hand. This was the first x-ray ever taken and was to become one of the most famous images in science, not only because it led to the routine and widespread use of x-rays in medicine, dentistry, and many other areas, but also because it was a major step in determining the structure of the atom and discovery of subatomic particles. In 1901 Röntgen became the very first recipient of the Nobel Prize for physics for his discovery of x-rays.

On January 7, 1896, two months after Röntgen discovered x-rays, the great French mathematician Henri Poincaré was sent x-ray images and was so amazed that he passed them on to two doctors and asked if they could duplicate Röntgen's work. On January 23, 1896, those Parisian physicians presented papers on the new phenomenon to the French Academy of Sciences. The physicist Antoine-Henri Becquerel (1852–1908) was in the audience when the x-ray photos were shown. Becquerel, the third Becquerel in a row to occupy the physics chair at the Natural History Museum in Paris, was particularly interested in this subject because he'd been studying phosphorescence produced by a number of different types of rocks that glowed in the dark and now wanted to test his hypothesis that there was some fundamental connection between x-rays and light. He spent the next several weeks experimenting with different types of materials, and by February 24, 1896, he was confident enough about his conclusion that he presented a paper to the academy in which he reported that phosphorescence from the rocks exposed to ultraviolet light from sunlight also emits the mysterious x-rays. However, when the sensitive photographic plates later recorded an image despite being stored in a dark drawer, he realized he was wrong. Becquerel promptly figured out it was the uranium in a piece of potassium uranium sulfate in the drawer that was spitting out some type of radiation all by itself—that is, without exposure to sunlight. Becquerel continued researching this phenomenon and published seven scientific papers on the subject in 1896.

As word spread around the world, physicists expanded their investigation of x-rays to the nature of the emanations from uranium. The source and nature of these emanations—labeled "alpha particles" or "alpha rays"—were completely unknown. The idea that a rock could beam out rays strong enough to penetrate the photographic plates was contrary to everything scientists had believed, particularly the law of conservation of energy, which held (at the time) that energy could never be created or destroyed. In contrast to the x-rays that Röntgen had created by electricity and the cathode ray tube, Becquerel's production of radiation seemed to come from *no place at all*. Thus physicists pursued the search within the uranium atom to try to determine the source of the alpha particles.

In early 1897, Joseph John Thomson (1856–1940), director of the Cavendish Laboratory in Cambridge, England, the most prominent physics lab in Europe and perhaps the world at the time, was the first to discover that at least one type of the radiation being emitted from Röntgen's cathode ray tube consisted of a stream of small particles (which he called corpuscles, now known as electrons), each of which contained a negative charge. On April 29, 1897, Thomson presented a paper on this theory to the Royal Society, claiming that these particles were constituents of matter and were smaller than any known object, including the atom. This prompted an uproar because the scientific community did not yet believe there could be anything smaller than an atom. Thomson speculated that Becquerel's alpha particles from the uranium were also electrons, but there was no evidence to support that theory. At the same time as Thomson was conducting his experiments, Heinrich Hertz (1857–1894), a professor at Bonn, Germany, reported that metal such as zinc that had been given a negative charge also gives up those charges when exposed to ultraviolet light. As a result of further work by Thomson, Hertz, and others, scientists became convinced of the existence of electrons.

The discovery of electrons then led to the proposition by some physicists that if there were negatively charged bits of matter in atoms, there must be positively charged matter to balance them out. But with little information on which to base a theory, their thinking was vague about such a positive counterforce, and no one had yet suggested a clear or convincing image or description of the atom. Therefore, at the end of 1897 scientists had varying concepts of the atom: a "buzzing hive of thousands of electrons somehow held together by a positively charged glue" or "a spongy, doughy blob," or "a plum pudding," as J. J. Thomson suggested, with the electrons being the plums embedded in the center of the atom, the "nucleus."

Marie Curie Leads Way in Identifying Alpha Particles and Proving Existence of Radioactive Elements

Acid-Stained Hands and Leaky Roofs

In 1891, Marie Curie (1867–1934) (born Maria Skłodowska) had left her native Poland and arrived at Paris's Gare du Nord train station. The Eiffel Tower, the centerpiece of the 1889 World's Fair, loomed over the city as a symbol of modern technology, along with electric lamps and internal combustion vehicles on the great boulevards. She'd come there to live near her sister Bronisława—a doctor, who'd made it through school partly with Marie's financial support from her hard-earned and meager savings—reluctantly leaving their father and the rest of the family in Warsaw. Marie's hands soon became acid-stained for life as she performed chemistry experiments in Professor Gabriel Lippmann's lab at the Sorbonne in the hopes of a career in math and science.

In 1896 she received her certificate to teach secondary education to girls, and in the following year decided to seek a PhD in science, an achievement that no European woman had ever accomplished. Having been fascinated by the recent discoveries of x-rays by Röntgen and alpha particles by Becquerel, the thesis subject she chose was the study of alpha particles. *What was the source of the energy that darkened Becquerel's photographic plates? What were the rays emanating from uranium?* Perhaps these were a stream of electrons, as Thomson had suggested, but nothing was known for certain about these mysterious units of matter. She selected this topic because the field was so new and could be pursued by doing experimental work in the laboratory, which she loved (as opposed to wading through books and articles in the library), and because it was a narrow and well-defined objective. As physicists groped for more detail and definition in their concepts and theory of the atom, Marie Curie began a friendly rivalry with Becquerel and a physicist named Ernest Rutherford to be first to understand the source, nature, and behavior of the alpha particles that Becquerel had first discovered. This subject would dominate the rest of her life and forever tie her name to the history of nuclear physics.

Curie's work began in an unheated, damp, abandoned shed with a roof that leaked in the Paris downpours. Her first notes were dated December 16, 1897, when she began using the piezoelectric quartz electrometer invented by her husband, Pierre, and her brother-in-law, Jacques, a mineralogy professor. The piezoquartz electrometer was sensitive enough to measure the feeble electric current emanating from uranium and other substances. Pierre was head of the laboratory at the newly founded School of Industrial Physics and Chemistry in Paris, and worked closely with Marie on her research. However, it was Marie who first developed the theory that the emission of the rays must be a phenomenon coming from within the atoms of uranium itself. That is, the rays were part of the matter, unlike x-rays, which are now known to be a form of electromagnetic radiation, and not something absorbed and returned in the form of phosphorescence. She coined the term *radioactivity* from the Latin *radius* meaning "ray." This simple hypothesis was to become Marie Curie's most important contribution to science because it opened up the structure of the atom for others to build on, leading in the early twentieth century to the unfolding of complex atomic structure.

Remarkably, on April 12, 1898, only a few months after she began this research, with no prior sources to rely on, Marie issued her first report, "Radiation Emitted by Compounds of Uranium and Thorium." The paper was presented to the French Academy of Sciences on her behalf by Professor Lippmann, because papers could be presented only by academy members. On July 18, 1898, Marie and Pierre discovered the previously unknown element polonium (which she named after Poland), and on December 26 of that year, only a year after beginning to work on her thesis, the pair announced the existence of another new element, which they called radium. She determined that radium was a million times more radioactive than the uranium rock that had originally sparked all the research on radiation. Needless to say, Marie Curie earned her PhD in science. In 1903, she and Pierre, together with Henri Becquerel, received the Nobel Prize for physics for their discovery of radioactivity. Her work so transcended the origins of atomic theory that this was not to be her only Nobel. Remarkably, her second Nobel Prize was awarded in chemistry, unshared, in 1911.

Rutherford Develops the Modern Picture of the Atom

Gold Leaf and Alpha Particles

New Zealander Ernest Rutherford (1871–1937) came to England in 1895 as the first foreign research student ever admitted to the Cavendish. He left the Cavendish in 1898 to accept the physics chairmanship at McGill University in Montreal, where he continued his life's work studying alpha rays. During eighteen months of his nine years there, he worked with Frederick Soddy (1877–1956), a young chemist from Oxford, and discovered that elements that display radioactivity slowly change into other elements—further proof that the law of conservation of energy was not accurate and that chemical elements were not immutable. For example, the radioactive element thorium decayed into helium and other elements. This work, published in 1902 in *Philosophical Magazine*, was the most significant contribution to the understanding of radioactivity since the Curies' discovery of radium four years earlier.

In 1903, as some scientists were still struggling with Thomson's plum pudding model of the atom, others suggested that atoms are made of varying numbers of one type of constituent called "dynamids." Then in 1904, Japanese physicist Hantaro Nagaoka proposed that the electrons were orbiting in a "saturnian ring"—a positive sphere was envisioned with a halo of mutually repellant electrons resembling the planet Saturn. While these theories of the structure of the atom were being considered, scientists determined in 1906 that the x-ray was a form of electromagnetic radiation of extremely short wavelength. Though Röntgen's discovery of x-rays initially led to research on the atom, and there would later turn out to be a direct relationship between atomic structure and x-rays, most questions of the day focused on *subatomic particles* like electrons and alpha particles, not electromagnetic *radiation* like x-rays, light, or radio waves.

In 1907, Rutherford returned to a post at the University of Manchester in England and continued an ongoing debate he was having with Becquerel on the behavior of the alpha particles that were spewed out of radioactive material. When Becquerel reported in his experiments that alpha particles appeared to be bouncing off air molecules in their path, Rutherford began his own work to find out how that occurs. With help from his assistants, including the German physicist Hans Geiger (1882–1945)—who later invented the Geiger counter, the first successful detector of alpha particles— Rutherford determined that 1 alpha particle in 8,000 rebounded when it slammed into a sheet of gold leaf.

Finally, in late November 1911 Rutherford determined that to deflect alpha particles, most of the atom would have to be concentrated into a tiny charged core in its center (the nucleus). He presented this theory to a session of the Manchester Literary and Philosophical Society on March 7, 1912. The deflection of alpha particles, proposed Rutherford, must be caused by something very small and very hard in the atoms of the material against which such particles collide, such as the sheet of gold with which Rutherford had been experimenting. He rejected Thomson's plum pudding model of the atom because electrons wouldn't bounce off it and instead described the

FIGURE 5-1

Rutherford's Apparatus for Studying Elements Under Alpha Bombardment

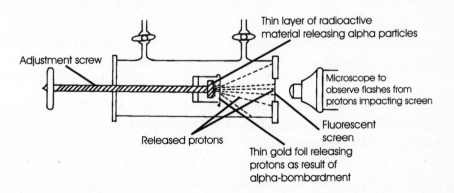

atom as "a central electric charge concentrated at a point," referring to the nucleus, with "*a uniform spherical distribution of opposite electricity,*" referring to the electrons. It doesn't matter which charge is called positive and which negative; the point is that they're opposite charges. The "uniform spherical distribution" of electrons is traditionally called negative.

Rutherford's public statement of his description marks the beginning of the modern understanding of the structure of the atom. He was first to the finish line, ahead of Curie and Becquerel. We now know that the alpha particles that Becquerel first discovered are the nuclei of ordinary helium atoms and are catapulted from decaying radioactive atoms at speeds over 10,000 miles per second. For Rutherford's picture of the atom, he received many honors and was knighted in 1914, but he was never awarded the Nobel Prize for physics.

Although Rutherford's work provided the basis for the modern concept of the atom, serious questions remained. A number of physicists, including Rutherford himself, soon realized that if he were correct, there would have to be a reason that the orbiting electrons would not be sucked into the positively charged nucleus. In other words, Rutherford's theory was groundbreaking, but his atom was both mechanically and electromagnetically unstable. Thus there began a concerted effort to find some type of balancing mechanism within the atom.

Bohr Determines That Fixed Quantities of Energy Allow Electrons to Maintain Their Orbits

Honeymoon Missed . . . Collecting Tolls in Light Quanta

A native of Denmark, Niels Bohr (1885–1962) came to England on a scholarship in 1911 to work with J. J. Thomson at the Cavendish for one year. He'd studied Thomson's plum pudding nucleus and, like Rutherford, thought it could not possibly be correct. Shortly after arriving, Bohr had his first opportunity to engage Thomson in a discussion about his concept of the nucleus, but the Dane's poor grasp of the English language limited his ability to express himself, and he blurted out to the famous scientist, "This is incorrect!" Because of their differing concepts of the atom, Bohr's stay at the

Cavendish was not pleasant and he arranged to begin working for Rutherford at Manchester in March, 1912, the same month that Rutherford presented his now-famous paper on the atom's structure.

His association with Rutherford proved fruitful because the two shared a similar concept of the relationship between the electrons and the nucleus, including the fact that until someone adequately addressed the *instability* of the atom, no one would accept the picture that Rutherford had proposed. In the late spring of 1912, just a few months after associating with Rutherford, Bohr suggested that a theory about light developed by Max Planck and Albert Einstein might apply to the atom. Planck and Einstein had said that light consists of discrete amounts of energy called *quanta*. Bohr then speculated that *quantization* might be a fundamental property of *all* energy. If so, this could provide an explanation for the atom's stability. *Could fixed quantities of energy be related to fixed electron orbits around the nucleus?* Bohr set out to answer this question.

He returned to Copenhagen from Manchester in the summer of 1912 and married his fiancée, Margrethe Nørlund. They canceled the honeymoon they'd planned so Bohr could spend the time developing his theory. Seven months later, in February 1913, Bohr was still struggling with how to describe this theory from both a conceptual and mathematical standpoint when a friend suggested that he look into a formula developed by Swiss schoolteacher Johann Balmer (1825–1898), which gives the frequency of light emitted by atoms. Bohr immediately realized that a combination of the Planck/Einstein quanta with this formula would provide the information he needed to describe the behavior of electrons orbiting the nucleus. His application of the formula to quanta demonstrated that:

- Electrons must exist in fixed arrangements, like satellites orbiting a globe at specified distances away from it.

- An electron leaping from one orbit to another results in an increase or release of a "quanta" of energy.

An atom does not emit radiation while it is in one of its stable states, but only when it makes a transition from one state to another. That energy is in the form of electromagnetic radiation, some of which can be seen in the form of light. As explained by Robert Crease and Charles Mann in *The Second Creation*, "Quanta of light are the tolls collected or paid by electrons as they jump about their permitted places in the atom. The spectral lines described by Balmer's formula are a set of sub-atomic hops, skips, and jumps, the jitterbug moves of hydrogen's lonely electron." The atom neither absorbs nor emits radiation continuously but only in "quantum leaps," a term commonly used outside of physics today that owes its origin to Niels Bohr's theory.

On March 6, 1913, almost one year to the day after Rutherford had described the atom as a central electric charge with a uniform spherical distribution of opposite electricity, his assistant and colleague Bohr sent a paper to Rutherford, describing how his new model for atomic structure explained the hydrogen spectrum. Bohr published his own theory in the *Philosophical Magazine* issues of July, September, and November 1913, and provided the solution to the major problem left open by Rutherford. The Rutherford-Bohr concept of the atom was born.

Bohr's work was confirmed later that year by the English physicist Henry Moseley (1887–1915) through a series of experiments that proved that the high-energy light given off when electrons move from a high to a low energy state was in the form of the x-rays originally discovered by Röntgen in 1895—thus the earlier-mentioned connection between x-rays and electrons (page 75)—and that there's a mathematical correlation (known as Moseley's

FIGURE 5-2
Rutherford-Bohr Model of Carbon Atom

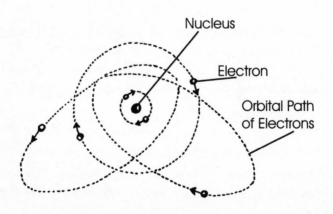

law) between the x-ray energy of each element and its atomic number. If the frequency and thus the energy of the x-rays are known, the charge on the nucleus can be determined. This measure of x-ray energy, which is specific for each of the elements from which it emanates, is commonly used today to detect elemental "fingerprints" in any substance, such as a fluid, rock, or microscopic particle of unknown composition.

As mentioned in Chapter 4, the nineteenth-century chemists Avogadro and Cannizzaro figured out that atoms bind together into "firm unions" or molecules. However, it was not until the Rutherford-Bohr model of the atom was developed that we understood exactly how this chemical bonding occurs. The association of atoms into molecules occurs when there are "unpaired" electrons in the atom's outer shell or orbit. As a result of the tendency of electrons to distribute themselves among atoms so the total energy of a group of atoms is lower than the sum of the component atoms, electrons are *shared* by atoms and thereby form molecules. The number of unpaired electrons in the outer shell is called the *valence* number of the atom, and the bonding or sharing of electrons is called *covalency*.

To help understand the scale of this central component of all matter, envision a single atom blown up to the size of the Superdome in New Orleans

FIGURE 5-3
Molecules

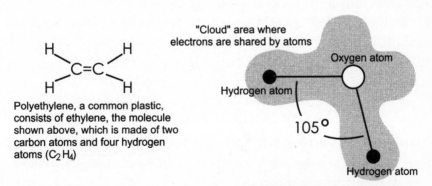

Polyethylene, a common plastic, consists of ethylene, the molecule shown above, which is made of two carbon atoms and four hydrogen atoms (C_2H_4)

"Cloud" area where electrons are shared by atoms

Oxygen atom

Hydrogen atom

$105°$

Hydrogen atom

Water (H_2O) contains a central atom of oxygen flanked by two smaller atoms of hydrogen (which seats more than 75,000 people). The atomic nucleus would be the size of a pearl in the center of the structure, with the negatively charged electrons orbiting at a speed of thousands of miles per second at the outer perimeter of the building. Scaled down somewhat, if the nucleus were enlarged to the size of the period at the end of this sentence, the electrons revolving around it would be about seventy-five feet away, with that enormous amount of space between the dot and those electrons being totally empty. The atom is virtually empty. Its size is only a description of the space, or in the physicists' vernacular, the "cloud" through which its electrons vibrate and pass at that great speed. The diameter of the nucleus is about one-hundred-thousandth the diameter of the whole atom. The volume of the nucleus is one-trillionth of the whole atom. To put this in another perspective, if the Earth were enlarged to the incredible size of *93 million miles in diameter* (the distance from here to the sun), the nucleus of even the largest atom, increased proportionately, would still be microscopic. The atom and its basic characteristics seem so bizarre and foreign to everyday experience that it is difficult to conceptualize. This fundamental form of matter—fleeting, transient, ephemeral, surreal—is like an illusion. Yet the manifestation of the electrical charges exhibited by the elements is our tangible reality.

In 1921, the University of Copenhagen created the Institute of Theoretical Physics for Niels Bohr, where he served as director for the remainder of his life. The institute soon became the world capital of atomic physics, attracting leading researchers and lecturers from the world over. In 1922, Bohr received the Nobel Prize for physics for his work on atomic theory.

Following publication of Bohr's quantum theory, hundreds of theoretical and experimental physicists at the Cavendish and later at Bohr's institute delved deeper and deeper into the structure and workings of the atom—like archaeologists carefully chipping layers from a delicate fossil. Theories were constructed, then thrown into the worldwide academic arena to see if they could withstand laboratory experiments and critical analysis. Fierce intellectual battles ensued among the greatest names in physics, played out at laboratories, in correspondence, and in the pages of scientific journals. *Is the electron a wave or a particle? What causes electrons to change orbits? What forces keep the nucleus intact and thus matter stable? Is there another particle in the nucleus?* During the 1920s a new physical interpretation of the atom was developed, called quantum mechanics, which linked such physical principles to Bohr's and others' mathematical descriptions.

One major aspect of the research performed in the succeeding years was directed at a large uncharged (neutral) particle suspected to be present in the nucleus. Rutherford had predicted the existence of such "neutrons" as early as 1920 and conducted a series of experiments at Cambridge in an attempt to find them. However, these experiments were ultimately inconclusive, and his idea was pushed aside. But in 1932, Rutherford's former student James Chadwick (1891–1974), who was working at Cambridge University and was well aware of Rutherford's prediction, confirmed the existence of the neutron in the nucleus one year after Irene Curie and Frederic Joliot, the daughter and son-in-law of Marie Curie, handed Chadwick a decisive clue. They reported the discovery of a new type of high energy but *uncharged* "radiation" made by bombarding beryllium with alpha particles. Curie and Joliot mistook this as gamma rays, thus leaving Chadwick to discover the elusive neutron, for which he won the Nobel Prize for physics in 1935. Chadwick's discovery of the neutron not only completed the picture of the atom's basic structure consisting of three "pieces," it would also become an essential step in harnessing and releasing the enormous power hidden in the nucleus.

On July 4, 1934, Marie Curie died at the age of sixty-seven, a victim of nearly four decades of overexposure to radioactive elements. By this time, her legacy was evident all around her, as the discoveries she'd made rippled forward with increasing momentum into the 1930s and as knowledge of the atom unfolded. The discovery of the neutron, together with several other discoveries that took place in the 1930s, dramatically and permanently transformed this pure laboratory science into a field with practical application. The world was rapidly approaching the nuclear age.

Bohr's institute remained a focal point for theoretical physics through the 1930s and played a pivotal role in understanding how to split the atom, as we'll see in Chapter 6. When World War II erupted in 1939, Denmark was overrun and occupied by the Germans, and in 1943, in fear for his life because of his Jewish ancestry, Bohr escaped to England, then worked for two years on the Manhattan Project in Los Alamos, New Mexico, as part of a British research group.

The work conducted by physicists since the time Bohr accurately described the atom in 1913 fills hundreds of biographies and scores of texts with material well beyond the scope of this book. Not only have researchers "synthesized" new chemical elements but they've also broken down the three constituent subatomic parts of the atom into even smaller particles, as described later. As for new elements, we're up to number 118 (ununoctium; Uuo) as of this writing (see page 67); thus there are twenty elements beyond californium (98), the heaviest naturally occurring one. For example, copernicium (element 112) was formed in the lab by slamming zinc ions into a lead nucleus. However, like the other artificially created heavy substances, this new element is highly unstable and decays in milliseconds. All such elements with atomic numbers greater than 98 (californium) were discovered as synthetic products of particle accelerators at nuclear research laboratories or as debris from nuclear explosions. They are highly radioactive and have been synthesized in minute quantities, usually only a few atoms at a time, as was ununoctium.

The further breakdown of electrons, protons, and neutrons has been accomplished by the United States' spectacularly high-energy atom smasher, the Tevatron Collider, at the Fermi National Accelerator Laboratory in Batavia, Illinois. For twenty-five years, the Tevatron was the world's most powerful collider, but in 2009, it was replaced at the top by Europe's Large Hadron Collider (LHC) located near Geneva, Switzerland. At these facilities, researchers have discovered a proliferation of subatomic particles. There are actually *hundreds of such particles*, and the foremost achievement of the Tevatron was to reveal a particle called the "top quark," so-called because it has the most mass of all the elementary particles. Quarks join in threes to form protons and neutrons that make up the atomic nuclei, as discussed earlier. Using the colliders, physicists accelerate a single beam of protons into fixed elemental targets, producing multiple other particles, such as quarks, leptons, muons, gluons, and bosons, the functions of which fill the pages of modern physics texts and are well beyond the scope of this book. The majority of these particles decay into others within tiny fractions of a second. This fundamental structure of matter into six quarks and six leptons is called the *standard model*. Included in this model is the elusive Higgs boson particle, which was predicted by Peter Higgs in 1964 to exist according to the standard model and is responsible for giving the elemental particles their masses. Referred to in a book as "the God particle," the Higgs boson appears central to finally understanding the structure of matter. Physicists using the Tevatron were unsuccessful in finding it, but researchers employing the advanced atom-smashing power of the LHC have finally produced astounding evidence for the presence of the massive Higgs boson. This is the first new fundamental particle physicists have identified since 1995, and they state that they have "found the last piece in their standard model of fundamental particles and forces." The Higgs is more like an electric "field" than an actual particle, and when electrons and quarks go zipping through the so-called Higgs field, it produces a "drag" or inertial effect on those subatomic particles, imparting to them their mass. Stuart Raby from Ohio State University said at the announcement of the Higgs boson's discovery on July 4, 2012, that "The standard model is framed on having a Higgs Boson. If the Higgs weren't there, the whole theory would have to be rethought." Some scientists are suggesting that there will be no further subatomic particles to find, whereas others believe an even wider door has been opened by the Higgs boson. As with other great scientific discoveries, this will take time to be confirmed or refuted.

Was Anaxagoras right? Is matter infinitely divisible? Although that question can't yet be answered for sure, it is known with certainty that the subatomic particles discovered so far have specific attributes: electric charge, mass, specified quanta of energy, angular momentum, spin (direction of rotation), and velocity of such rotation.

Ernest Rutherford predicted that "the forces holding the nucleus together will not be found in this generation or the next . . . or for many hundreds of years, because the constitution of the atom is, of course, the great problem that lies at the base of all physics and chemistry." Though we will not delve further into the story of the creation of new atoms or the research into quarks, leptons, or angular momentum, we will explore how physicists came to understand "the forces holding the nucleus together." In the course of their research into the atom, they discovered something so incredible and with such far-reaching ramifications to all of us that it must be part of every discussion of the structure of the atom: the awesome power concealed and contained within the atomic nucleus. The bomb and nuclear power, with their mixed legacy of fear and wonder, destruction and promise.

Therefore, the next chapter looks at how scientists identified and harnessed the unbelievably immense force and energy in this surreal, illusory, impossible, invisible, texture-less, electrically charged "pearl" suspended in the central point of the Superdome.

Chapter Six
THE CRACK OF DOOM

I firmly believe that before many centuries more, science will be the master of man. The engines he will have invented will be beyond his strength to control. Some day science shall have the existence of mankind in its power, and the human race commit suicide by blowing up the world.
—Henry Adams (1862)

The atom stands as a monument to the wisdom of the Human Race. One day it may stand as a tombstone to its folly.
—J. G. Feinberg, *The Story of Atomic Theory and Atomic Energy* (1960)

The world is very different now. For man holds in his mortal hands the power to abolish all forms of human poverty and all forms of human life.
—John F. Kennedy, Inaugural Address (1961)

The free world rejoiced when the Berlin Wall came tumbling down in 1989, and we gave a collective sigh of relief on September 12, 1990, when the cold war came to a formal end with the Treaty on the Final Settlement with Respect to Germany. Signed by the United States, France, England, and the Soviet Union, this agreement allowed German reunification. We further relaxed in July 1991 when the START Treaty (Strategic Arms Reduction Talks) was signed and implemented, signaling the beginning of the dismantling of tens of thousands of nuclear warheads. These developments reduced the immediate threat of nuclear holocaust. However, as mere political changes are of questionable strength and duration, agreements on paper can never fully eliminate the uncertainty and anxiety concerning nuclear weapons and their potential to destroy civilization. Truces among nations that have conflicting political systems and otherwise compete in this world are not institutions. They are, at best, a temporary and highly fragile state of affairs. Indeed, that fifty-year preoccupation has been replaced to a significant extent in recent years with another fear: that less stable governments with existing or developing nuclear capabilities, North Korea, Pakistan, and Iran, for example, will act irresponsibly or will use a nuclear threat to the detriment of Western democracies or the international community in general.

Thus the threat of nuclear winter still hovers like a dark cloud. Ever since the secret of the mushroom cloud escaped from the atom's nucleus, the destinies of the atomic bomb and the human race became intertwined forever. The specter of that cloud will be with us as long as there is political uncertainty in the world. As long as there are nations, history has proven that civilizations will arise and vanish, and super powers will be born and die in the centuries and millennia ahead. In a hundred years, who will threaten to use the atomic bomb? As Theodore Zeldin said in *An Intimate History of Humanity*:

> If a film were to be made compressing into a couple of hours all that . . . happened . . . the world would look like the moon, grey and desolate, remarkable only because there are a few craters on it. The craters are civilizations—thirty-four major ones so far—each exploding then dying out, having briefly lit up parts of the globe, but never all of it; some last for a few hundred years, others for a couple of thousand. . . . All civilizations, so far, have decayed and died, however magnificent they have been in their glory, however difficult it is to believe that they can vanish.

We came to discover and unleash the secret symbolized by the mushroom cloud when our quest to fully understand the structure and function of the atom intersected with the ever-changing political and military history of the world.

Scientists Tie the Sun's Energy to the Atom's Nucleus

Missing Mass and the Key to All Life

Nineteenth-century scientists accepted the "contractive hypothesis" of the Scottish physicist William Thomson (1824–1907) (known as Lord Kelvin), which held that the sun obtains its energy supply from a slow but steady shrinking of its giant body. However, this theory accounted for only enough energy to support the sun's radiation for about 20 million years. Thus when biologists, geologists, and paleontologists discovered that evolution of life on Earth had begun well before 20 million years ago, Kelvin's theory was abandoned.

Henri Becquerel's 1896 discovery of spontaneous energy emanating from radioactive materials provided the first clue to the sun's immense energy, but it was not until Einstein introduced his famous formula ($E = mc^2$) that the true relationship between matter and energy, and the true source of the sun's light and heat, was understood. $E = mc^2$ means *energy equals mass times the speed of light squared*. With this equation, Einstein was proposing that matter ("mass") and energy are both *destructible*. In fact, one can be *converted* into the other. This was directly contrary to the supposedly inviolate Law of the Conservation of Energy—formulated in 1847 by the German physicist Hermann Ludwig Ferdinand von Helmholtz—which stated that neither matter nor energy could be destroyed or created. Einstein's theory raised the possibility that the sun's matter was being destroyed in some way and was being converted into energy in the form of heat and light.

Mass is the *quantity of matter* in the particular object, as mentioned in the discussion of Newton's laws of physics. That is, it reflects the actual number of protons and neutrons in the atoms that make up the object, as we saw in Chapter 5. Unlike the *weight* of the object, mass is not subject to change and doesn't depend on the gravitational pull of the Earth or other body. Therefore, physics uses this universal measure of mass. On a larger physical scale, mass is determined by the amount of force required to produce a given acceleration, as suggested by Newton's Second Law of Motion (acceleration is directly proportional to the force).

According to Einstein's equation, the amount of energy stored in the mass of any matter is incredibly huge. As stated, the energy (E) equals the particular unit of mass (m) multiplied by the speed of light (c) (186,000 miles per second) squared. In contrast to atomic energy, if you were to burn one-fifteenth of an ounce of hydrogen gas, it would produce enough energy for a 100-watt light bulb to burn for about forty minutes. But if you convert the mass (that is, all the atoms) of the same amount of hydrogen gas into energy, it would result in enough energy to power the same light bulb for *56,000 years*. Similarly, 1 pound of the radioactive element uranium yields as much energy as 3 million pounds of coal. Burning is simply a chemical reaction in which energy that is stored in the molecular structure is released. In other words, burning is a realignment of the *chemical* bonds between electrons, not a release of energy stored in the atom's nucleus.

Scientists came to realize that the sun was creating its energy by *fusing* hydrogen atoms together to form helium atoms—four hydrogen atoms fuse together to form one helium atom. When this happens, as is occurring continuously in every star throughout the universe, mass is converted into energy. Put simply, when four hydrogen atoms combine to make one helium atom, that resulting helium atom has slightly *less* than four times the mass of one hydrogen atom. The "missing mass" has been converted to energy, which we see (and depend on to exist) in the form of light and heat. This combination or fusion of hydrogen atoms into helium atoms is due to the incredible heat of the sun's core—18,000,000 to 36,000,000°F. This phenomenon, the conversion of mass to energy, is of the ultimate significance. It is the single factor that accounts for all life on this planet.

Ironically, Newton actually foreshadowed Einstein's equation in his famous book *Opticks* when he wrote, "Are not gross Bodies and Light convertible into one another, and may not Bodies receive much of their Activity from the particles of Light which enter their Composition?" However, the idea that this energy could somehow be captured and used, as suggested by Einstein's 1905 proposal, remained a curious and intriguing speculation for twenty-seven years because there was no way to test it. That is, there was no way to duplicate those enormous temperatures. In fact, soon after Einstein proposed that $E = mc^2$, he noted, "There is not the slightest indication that energy will ever be obtainable."

Scientists Determine That the Energy of the Atom's Nucleus Is Obtainable

Understanding the Nuclear Fluid

In 1928, the Russian-born nuclear physicist George Gamow (1904–1968) developed the first quantum theory of radioactivity, which earned him a fellowship at Niels Bohr's Copenhagen Institute of Theoretical Physics. While there (1928–1930) he formulated the "liquid droplet" model of an atomic nucleus, according to which different nuclei are considered the minute droplets of a universal "nuclear fluid." He based his theory on the assumption that the forces acting between the constituent parts of an atomic nucleus are similar to those acting between the molecules of an ordinary liquid. Gamow noted that the sphere is the geometrical figure that possesses the smallest surface for a given total volume. If there is no external force acting on a drop of liquid (such as gravity pulling it down onto a flat surface), the natural surface tension or surface energy will work in drawing and keeping the constituent parts together in the form of a sphere. Likewise, he surmised that it would take "extra" energy to overcome the force holding the nucleus of an atom together, the so-called strong nuclear force. He further pointed out that there is an energy opposing that surface tension type of effect—namely, the repulsion among and between the positively charged protons in the nucleus. The nucleus is a cohesive and tightly closed body of particles despite the natural tendency of the protons to split apart like the positive ends of two magnets. This indicates there is a natural equilibrium in every atomic nucleus. In other words, because the nuclei do not spontaneously split into smaller pieces or fuse together with other nuclei, Gamow reasoned that the strength of the surface tension of the "nuclear droplet" *exceeds* the natural internal repulsion force of the nucleus.

The electric repulsion force among the protons and the strong nuclear force holding the nucleus intact each have a specific mathematical value. By applying those values to the nucleus, Gamow demonstrated *theoretically* that splitting the nuclear droplet into two halves would result in an *increase* in the total surface energy, while the electric repulsion force among the protons

FIGURE 6-1

Nuclear "Droplet" with Forces

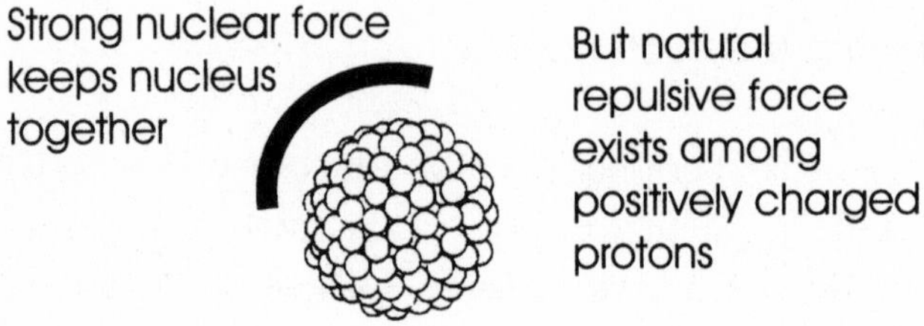

Strong nuclear force
keeps nucleus
together

But natural
repulsive force
exists among
positively charged
protons

would be *reduced*. Thus nuclear physicists began to better understand what happens to that reduction or release of energy from the nucleus when fusion or fission takes place and began to seriously consider that it could be captured and put to some practical use.

As discussed in Chapters 4 and 5, there are ninety-eight naturally occurring elements of increasingly larger atomic weight and mass. The nuclei of hydrogen, helium, and some other light elements were formed relatively soon after the Big Bang when the universe was extremely hot and dense, while other nuclei and atoms were formed as a result of stars' fusion reactions and as a result of stars exploding at the end of their normal life cycles. In the present cool universe, all these elements are in a very stable state. In Chapter 5, we saw how electrons transfer freely to the outer orbit of adjacent atoms when they form molecules. In contrast to this phenomenon, the strong nuclear force holding the nucleus together is much stronger and more complicated. The *strong nuclear force* is one of the four basic forces of nature, along with *gravity*, the *electromagnetic force*, and the *weak nuclear force*, which connects the orbiting electrons to the nucleus and is the force responsible for holding together atoms that combine to form molecules. Gravity and electromagnetism act over great distances across the universe, whereas the strong force can be detected only in a laboratory because it operates on a tiny scale in the atom's nucleus, binding protons and neutrons together.

Despite the extremely localized effect of the strong nuclear force, it would take a large amount of energy to destabilize the nucleus and create fusion or fission—much more "activation" energy than is needed to heat an ignitable substance before it will burn or charge a high explosive in order to detonate it. For fusion to occur, nuclei must be propelled toward each other to overcome their inherently repulsive forces. For fission to take place, a violent collision is usually needed to break up a nucleus, with the exception of naturally radioactive atoms, which undergo spontaneous fission. Such radioactive atoms contribute substantially to the internal heat of the Earth and decay at rates so statistically reliable that they are used as natural clocks and have invaluable application in geology to determine the age of rocks and other substances on Earth (as discussed on pages 65–66).

By the early 1930s, based largely on Gamow's nuclear droplet model, scientists realized that although it likely would be impossible to duplicate the sun's *fusion* process caused by its enormous interior temperature, they might be able to harness the huge amount of energy predicted by Einstein's formula through *fission*—that is, *splitting the atom's nucleus.* Harnessing any appreciable amount of energy through fission is more complicated than merely knocking off part of the nucleus. Henri Becquerel and Ernest Rutherford had been chipping protons off the nuclei of atoms since early in the century by bombarding them with alpha particles from a radioactive source and their labs didn't blow up. Similarly, in 1919, Rutherford had succeeded in splitting nitrogen atoms into hydrogen and oxygen atoms by bombarding them with alpha particles. As we entered the 1930s, physicists pursued the idea of somehow striking enough atomic nuclei with such force that fission would take place on a scale sufficient to release and they hoped capture the energy escaping from the nuclear droplet.

The Nucleus Is Split and the World Enters the Atomic Age

Escape from Italy

With the deeper understanding of the forces holding the nucleus together, scientists realized they would need a device that would artificially project protons at nuclei with great speed. They knew it would be very difficult for such protons to penetrate the natural barrier of electric repulsion resulting from the protons in the nucleus. Such a successful violent collision first occurred in 1932 when the British physicist Sir John D. Cockcroft (1897–1967) and the Irish physicist Ernest Walton (1903–1995), working with Ernest Rutherford at Cambridge's Cavendish Laboratory, created the first atom smasher, which became known as the Cockcroft-Walton generator. They accomplished the first complete artificial splitting of an atomic nucleus (lithium nuclei) by bombarding it with protons and causing fission of lithium into two helium atoms. The energy gain was small, but it was the first demonstration of Einstein's prediction that energy would result from the conversion or loss of mass ($E = mc^2$), and might be captured and used.

Cockcroft and Walton were later awarded the Nobel Prize for physics. However, the Cockcroft-Walton generator and similar equipment that succeeded it were limited by the haphazard and unpredictable manner in which fission was accomplished. Without repetitive releases of energy from splitting nuclei, more energy is expended in accelerating the projectiles used to split the nucleus than is obtained from the fission itself.

As mentioned in Chapter 5, 1932 was also the year that James Chadwick discovered the existence of the neutron in the atom's nucleus—a particle with mass similar to the proton, but with no electrical charge. Scientists quickly realized that the ideal "atomic bullet" to split the nucleus would be the neutron, not the proton, because the neutron would not be electrically repelled by the positively charged nucleus. Thus another turning point in making atomic energy a reality occurred in 1935 when the Italian physicist Enrico Fermi (1901–1954) conceived the idea that (1) neutrons could be obtained from radioactive substances, (2) compounds rich in hydrogen (such as water) slowed down the neutrons being obtained from such radioactive compounds, and (3) when such neutrons were slowed, the likelihood that they'd be *captured* by an atomic nucleus *target* was increased. Fermi reasoned that when the large mass of the neutron impacts and binds with the target nucleus, the natural repulsive force between the protons would become greater than the surface tension of the nuclear droplet, thereby causing the nucleus to split apart. The incoming neutron could be sufficiently disruptive even to split a nucleus with many protons and large potential energy (say, the nucleus of a uranium atom) into two particles of approximately equal size, and release that energy.

In 1938, Fermi was awarded the Nobel Prize for physics for his work in neutron bombardment with slow neutrons. When Mussolini's fascist gov-

FIGURE 6-2
Fission and Chain Reaction

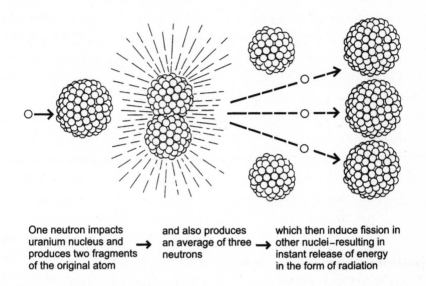

| One neutron impacts uranium nucleus and produces two fragments of the original atom | → | and also produces an average of three neutrons | → | which then induce fission in other nuclei–resulting in instant release of energy in the form of radiation |

ernment granted Fermi permission to go to Sweden to receive the award, Fermi and his family carried out their plan to permanently leave Italy. They soon settled in the United States where Fermi became a professor at Columbia University.

The Community of Nuclear Physicists Is Forced to Shift Its Focus

Yours Very Truly . . . Albert Einstein

As 1939 began, fission of a large nucleus into two halves and the release and harnessing of its enormous energy remained an abstract hope. If it could be achieved, nuclear physicists envisioned using atomic energy for peaceful industrial purposes for the good of humankind. Fission of a single atom or small number of atoms had become possible. However, scientists realized that they must achieve a continuous and prolonged series of proton bombardments in which the neutrons of such large nuclei actually become the projectiles to disrupt other nearby large nuclei (Figure 6-2). Without such a chain reaction, more energy would be used in accelerating the projectile than would be obtained from the fission, as Cockcroft and Walton had experienced. This was the year World War II broke out in Europe, and it is far from a coincidence that creation of the atomic bomb suddenly displaced the vision of harnessing atomic energy for peaceful use. Because the discovery of how to achieve fission through such a massive chain reaction was a direct outgrowth of the war, the bomb had its origin in 1939.

The Austrian mathematical physicist Lise Meitner (1878–1968) figured out how to achieve the fission and chain reaction of a heavy element, thereby becoming a key figure in the development of the atomic bomb. After conducting experiments in Germany similar to Fermi's neutron bombardment, Meitner and two of her colleagues realized they had actually split uranium atoms into several parts and created a chain reaction on a small scale. In January 1939, she secretly left Germany and went to Stockholm, where she and her nephew, Otto R. Frisch (1904–1979), a professor at the Institute of Theoretical Physics of the University of Copenhagen, working under Niels Bohr, then submitted a letter to the editor (titled "Disintegration of Uranium by Neutrons: New Types of Nuclear Reaction") in the most prestigious British science journal, *Nature*, explaining Meitner's experiment. Their report appeared in the February 11, 1939, issue and was first to use the term *fission* to describe the division of uranium into lighter elements:

> Upon splitting a uranium nucleus (which contains 54 more neutrons than protons), two new nuclei (such as barium atoms with 26 extra neutrons) would be formed, thereby resulting in a discharge of the neutron oversupply, thus further resulting in a chain reaction when those neutrons smash into the nuclei of other uranium atoms and continue the fission of the nuclei of such atoms.

Realizing the significance of Meitner's work, Fermi and Bohr engaged in studies and experiments in 1939 in the United States that further indicated the feasibility of a nuclear chain reaction. Later that year, Fermi and two other physicists at Columbia University, the Hungarians Leó Szilárd (1898–1964) and Eugene Wigner (1902–1995) brought the work on fission to Albert Einstein's attention. The U.S. government feared being drawn directly into the war and continued to support its allies against Hitler. Convinced the Nazis would be in possession of atomic power in the near future and would use it to subdue the rest of the world, Einstein agreed to sign a letter to President Roosevelt, drafted by Fermi, Szilárd, and Wigner. Although Einstein's previous work in physics had little to do with the direct discovery or the potential use of nuclear fission, he was asked to sign the letter because of his stature and credibility as a physicist. Unlike Einstein, the group of refugee scientists in the United States (that included Fermi, Szilard, Wigner, Gamow, and Edward Teller) had no direct channels of communication with the White House. The letter was dated August 2, 1939, but did not reach Roosevelt until October 11, 1939. Figure 6-3 shows a copy of that letter, which was to become one of the most important letters in the history of the world.

As a result of this letter, Roosevelt formed a committee charged with examining the prospects of the military use of atomic energy. After the committee reported to the president that it was possible to use such energy in atomic bombs, the U.S. Army and Navy made their first grant for atomic research in February 1940 in the amount of $6,000. Though a meager sum even by 1940 standards, this marked the first involvement of the U.S. military in atomic energy research. After signing and delivering the letter, Einstein had no further connection with research or development of atomic energy or the atomic bomb.

The Manhattan Project Is Initiated

Little Boy and Fat Man . . . 210,000 Dead

In August 1942, Roosevelt charged U.S. Army Brigadier General Leslie R. Groves with the responsibility of organizing the effort—with support from Great Britain and Canada—to harness nuclear energy for the military. This became known as the *Manhattan Project* because much of the research was being done by Enrico Fermi and others at Columbia University in Manhattan and because the army's Manhattan district office initially organized the project. In October 1942, Groves named J. Robert Oppenheimer (1904–1967) director of a newly formed secret weapons laboratory. The son of a German immigrant, Oppenheimer had graduated from Harvard in 1925, then did research in physics at the Cavendish Laboratory at the

FIGURE 6-3
Einstein's Letter to FDR

Albert Einstein
Old Grove Rd.
Nassau Point
Peconic, Long Island

August 2nd, 1939

F.D. Roosevelt,
President of the United States,
White House
Washington, D.C.

Sir:

Some recent work by E. Fermi and L. Szilard, which has been communicated to me in manuscript, leads me to expect that the element uranium may be turned into a new and important source of energy in the immediate future. Certain aspects of the situation which has arisen seem to call for watchfulness and, if necessary, quick action on the part of the Administration. I believe therefore that it is my duty to bring to your attention the following facts and recommendations:

In the course of the last four months it has been made probable - through the work of Joliot in France as well as Fermi and Szilard in America - that it may become possible to set up a nuclear chain reaction in a large mass of uranium, by which vast amounts of power and large quantities of new radium-like elements would be generated. Now it appears almost certain that this could be achieved in the immediate future.

This new phenomenon would also lead to the construction of bombs, and it is conceivable - though much less certain - that extremely powerful bombs of a new type may thus be constructed. A single bomb of this type, carried by boat and exploded in a port, might very well destroy the whole port together with some of the surrounding territory. However, such bombs might very well prove to be too heavy for transportation by air.

The United States has only very poor ores of uranium in moderate quantities. There is some good ore in Canada and the former Czechoslovakia, while the most important source of uranium is Belgian Congo.

In view of this situation you may think it desirable to have some permanent contact maintained between the Administration and the group of physicists working on chain reactions in America. One possible way of achieving this might be for you to entrust with this task a person who has your confidence and who could perhaps serve in an inofficial capacity. His task might comprise the following:

a) to approach Government Departments, keep them informed of the further development, and put forward recommendations for Government action, giving particular attention to the problem of securing a supply of uranium ore for the Untied States;

b) to speed up the experimental work, which is at present being carried on within the limits of the budgets of University laboratories, by providing funds, if such funds be required, through his contacts with private persons who are willing to make contributions for this cause, and perhaps also by obtaining the co-operation of industrial laboratories which have the necessary equipment.

I understand that Germany has actually stopped the sale of uranium from the Czechoslovakian mines which she has taken over. That she should have taken such early action might perhaps be understood on the ground that the son of the German Under-Secretary of State, von Weizsücker, is attached to the Kaisor-Wilhom-Inotitut in Berlin where some of the American work on uranium is now being repeated.

Yours very truly,

A. Einstein

(Albert Einstein)

University of Cambridge under Ernest Rutherford during the period of pioneering studies on atomic structure. He received his PhD in 1927 and began teaching physics at Berkeley. In 1939, when Germany invaded Poland and Einstein delivered the letter to President Roosevelt, Oppenheimer began working with the small circle of nuclear physicists who were aware of the potential of devising a nuclear bomb. His special area of expertise— separating uranium-235 from natural uranium, which is predominantly uranium-238—became a critical area of research. Under the cloak of secrecy, Oppenheimer assembled the best nuclear physicists available to work in various laboratories around the country and the new secret weapons laboratory located at a facility in the New Mexico desert, called the Los Alamos National Laboratory, to understand the implications of a chain reaction of heavy elements such as uranium to create fission. Oppenheimer's group included Enrico Fermi and these other individuals:

- **Hans Albrecht Bethe (1906–2005):** German-born physicist who worked with Fermi in Italy in 1931, emigrated to the United States in 1934, and headed the Theoretical Physics Division at Los Alamos. In 1939, Bethe had calculated the specific amount of energy created and the specific nuclear process involved in the sun's fusion.
- **Edward Teller (1908–2003):** Hungarian-born American nuclear physicist who studied at Niels Bohr's Institute in Copenhagen, collaborated with Gamow on understanding the nuclear decay process, worked on Fermi's team at the University of Chicago, and was one of the first recruited by Oppenheimer to go to Los Alamos.
- **Willard F. Libby (1908–1980):** American chemist who helped develop the method used in separating uranium-235 from uranium-238.
- **William G. Penney (1909–1991):** British nuclear physicist who led the team of British scientists working on the Manhattan Project.
- **Leó Szilárd (1898–1964):** Hungarian-born American physicist who was instrumental in persuading Einstein to sign the letter to FDR, developed the first method of separating uranium-235 from uranium-238, worked closely with Fermi on the 1942 chain reaction at the University of Chicago, and continued to work on the Manhattan Project until the end of the war.
- **Harold C. Urey (1893–1981):** American chemist who discovered deuterium (heavy hydrogen) in 1931, which advanced the development of

fission, and later directed the research at Columbia University during the Manhattan Project that provided fundamental information on separating uranium-235 from uranium-238.

Assigned the task of producing a controlled, self-sustaining nuclear chain reaction, Fermi and his colleagues pursued this work on an abandoned squash court beneath the stands at Stagg Field at the University of Chicago. His group had mathematically calculated that approximately 54 tons of uranium interspersed with about 400 tons of graphite would be the optimum volume of material to create such a chain reaction. On December 2, 1942, they were cautiously constructing this increasingly larger stack into a spherical shape when it came time for another test. As they slowly withdrew the cadmium control rods one by one, the meter that recorded neutron release and bombardment began to jump wildly. It reflected their success in obtaining a chain reaction—that is, the splitting of millions of uranium atoms. They reinserted the stable control rods to snuff out the activity. One of Fermi's assistants ran to the telephone to report the results to Oppenheimer. "The Italian navigator has just landed in the New World," he reported. This now-famous experiment represented the culmination of almost five decades of intense and continuous effort to fully understand the atom. Fermi's group was the first to demonstrate that Einstein's formula could have practical application, whether it be to provide needed energy in peacetime, or as a military weapon, or to "commit suicide by blowing up the world."

Once the chain reaction was moved from the realm of conjecture to the realm of the probable, secret plants were set up on a 70-square-mile area at Oak Ridge, Tennessee, and on an isolated 1,000-square-mile tract (named the Hanford Engineer Works) north of Pasco, Washington, on the banks of the Columbia River for access to its water to cool the enormous heat of the atomic piles to be constructed there. At Oak Ridge—where one of the buildings covered forty-four acres—the uranium-235 atom needed for fission could be stripped from its natural companion, the predominant uranium-238 material. At the equally massive Hanford complex, the experimental manufacture of plutonium-239 was to take place. Uranium-235 and plutonium-239 were earmarked as having the best potential for creating the explosive chain reaction necessary for a nuclear bomb. The population at Oak Ridge eventually reached 75,000, and at Hanford, 50,000.

The product resulting from the work by the people at Oak Ridge, Hanford, and other locations was intended to be used at the secret facility thirty-five miles northwest of Santa Fe, New Mexico, on the flat and desolate expanse of the Southwest desert called Los Alamos. The site had been chosen by Oppenheimer for its very remoteness in November 1942, a month after he'd been selected to head the Manhattan Project. At Los Alamos, he had combined the efforts of the great physicists he'd brought together, including Fermi and the others. The scientists at Los Alamos were given an awesome assignment: to design the actual bomb to be powered by the fissionable material being manufactured at Oak Ridge and Hanford.

One essential question they immediately faced was determining the amount of uranium-235 or plutonium in the bomb. That is, what is the "supercritical mass" of fissionable material? The time elapsing between the ejection of a neutron from a uranium atom that is being split and the resulting splitting of a second uranium atom is a *100-millionth of a second*. If too little uranium were used, the neutrons would escape from such material without creating the necessary chain reaction. But if too much uranium were gathered together closely, it would instantly explode and destroy Los Alamos and all 5,000 people there. If the Los Alamos team could solve this and the other enormous problems they faced in designing the bomb, its force—with each *pound* of uranium or plutonium exploding with the power of 8,000 tons of dynamite—would stagger the imagination . . . and perhaps end the war.

After about two years of work at Oak Ridge and Hanford, enough fissionable material was produced to enable the Los Alamos group to begin to physically assemble the first atomic bomb—a device about 6 feet long and 2 feet in diameter, weighing four tons, and containing what they had calculated to be the supercritical mass of plutonium-239. Thus about five and a half years after Roosevelt read Einstein's letter and after the expenditure of $2 billion and millions of man-hours, the countdown to test the physicists' invention began in the predawn hours of July 16, 1945, at the remote Alamogordo Air Base in New Mexico. The bomb was placed on top of a 100-foot steel tower surrounded by scientific monitoring equipment. Robert Oppenheimer and his group were in the control room six miles away from the bomb site, while other scientists and observers huddled in bunkers and behind shelters and other fortifications ten miles from the site. Time minus fifty minutes, time minus thirty, twenty, ten, and so on until the voice over the speaker shouted, *"Now!"* Suddenly, a ball of fire shot skyward, and the onlookers were flooded by a burst of golden, purple, gray, and blue light 100 times the intensity of the afternoon sun on a clear day. The detonation was the equivalent of 20,000 tons of dynamite. Thirty seconds into the furious blast, several observers standing behind a shelter were knocked down by the rush of air, even though they were ten miles away from the explosion. General Farrell, the chief military officer at Los Alamos, later wrote in his report to the War Department that the explosion was "unprecedented, magnificent, beautiful, stupendous and terrifying." The steel tower was completely vaporized and the surrounding desert sand for a radius of 800 yards fused to glass.

As the mushroom cloud rose to 40,000 feet in the air, the people at Alamogordo that fateful day witnessed the shape that would influence much of history for decades and perhaps centuries. They wrote the last chapter of the story that joined the past to the future, for this was the culmination of humankind's fascinating 2,400-year-long struggle to gain insight into the ultimate piece of matter—Leucippus's particle, the very beginning of all matter. This tale ended with an entirely unexpected and extremely frightening twist. Through an amazing application of human reasoning power, a line of outstanding scientists dating back to Robert Boyle and the other seventeenth- and eighteenth-century chemists not only figured out the potential within the atom but also made Einstein's formula a prophecy as the energy within the atom was harnessed and its fantastic secret released, to be with us now, in peace and in war, forever.

On August 6, 1945, three weeks after the test at Los Alamos, a single B-29 bomber, the *Enola Gay*, carried a bomb containing uranium-235 across the Pacific Ocean and released it over Hiroshima, Japan, at 8:15 in the morning. Nicknamed "Little Boy," the weapon detonated 1,900 feet above the ground, destroying two-thirds of the city and killing 140,000 of its 350,000 inhabitants. A second weapon, nicknamed "Fat Man," similar to the plutonium bomb tested at Los Alamos, was scheduled to be dropped on Kokura, Japan, on August 11, but was moved up two days to avoid bad weather predicted for the area. The B-29 *Bock's Car*, which carried this bomb, spent ten minutes over Kokura without locating its target through the clouds, then proceeded to its secondary target, the city of Nagasaki. At 11:02 a.m., Fat Man destroyed half the city and killed 70,000 of its 270,000 people.

PART THREE

The Principle of Relativity

In Focus

What could be more certain and absolute than the passage of time? At the age of sixteen, Albert Einstein posed a question to himself that led, ten years later, to his formulation of the special theory of relativity, which shows that the rate at which time passes is not certain or absolute. The concept of universal time must be replaced with a multitude of "personal" times. Time and space are no longer separate entities. They must be replaced with a four-dimensional entity: space-time.

Einstein's theory did not replace Newtonian physics. Special relativity applies only to an object moving at extreme speed in relation to another. Newton hadn't contemplated those conditions and didn't have the information about electromagnetic waves and the speed of light available to him. Ten years after Einstein discovered the special theory, he completed his development of the general theory of relativity—a new view of the source and effects of gravity throughout the universe. His impact in numerous other fields of science was also significant.

Einstein was one of the greatest intellects of all time. But he also immersed himself in the pressing social issues of his day, compassionately speaking out at every opportunity and making great personal sacrifices as he tried to make the world a more civil place. We will first look at the person behind the theory, then we will examine the theory that shook the very foundation of time.

Chapter Seven
PHILOSOPHER-SCIENTIST

How strange is the lot of us mortals! Each of us is here for a sojourn; for what purpose he knows not. . . . The ideals which have lighted my way, and time after time have given me new courage to face life cheerfully, have been Kindness, Beauty and Truth.

—Albert Einstein, "The World As I See It" (1931)

Albert Einstein was a philosopher. He also happened to be an extraordinary physicist. We know about him today because of his theories about space, time, and gravitation, but he was equally passionate about freedom, the value and purpose of human life, good and evil, education, religion, politics, government, and pacifism. Einstein wrote or spoke publicly on such issues more than 200 times between 1915 and his death in 1955, despite the fact that he shied away from his worldwide fame:

> Let every man be respected as an individual and no man idolized. It is
> an irony of fate that I myself have been the recipient of excessive admira-
> tion and reverence from my fellow-beings, through . . . no merit of my
> own. The cause of this may well be the desire . . . to understand the few
> ideas to which I have with my feeble powers attained.

Although Einstein devoted most of his life to science, he had a profound interest in many philosophical issues beginning at an early age. After stepping onto the world stage in 1915, he not only spoke and wrote about the many moral, political, and philosophical issues mentioned here but also corresponded with the famous Austrian psychiatrist Sigmund Freud about the love-hate instincts of humankind, engaged in extended philosophical dialogue on the nature of truth with the Hindu poet and mystic Rabindranath Tagore, and debated with many other thinkers of his day. He was also a gifted writer in the German language and a lover of music, and was most concerned about needless human suffering.

The following excerpts from Einstein's public statements and articles offer a glimpse into the rich body of ideas, ideals, opinions, and views of life espoused by this modest, brilliant, philosophical, and highly idealistic individual.

On the Military
In two weeks the sheeplike masses of any country can be worked up by the newspapers into such a state of excited fury that men are prepared to put on uniforms and kill and be killed, for the sake of the sordid ends of a few interested parties. Compulsory military service seems to me the most disgraceful symptom of that deficiency in personal dignity from which civilized mankind is suffering today. (Published in *Mein Weltbild*, 1934)

On Palestine
Jews and Arabs confront each other as opponents. . . . This state of affairs is unworthy of both. . . . A Privy Council is to be formed to which the Jews and Arabs shall each send four representatives. . . . These eight people are to meet once a week. They undertake not to espouse the sectional interests of their profession or nation but conscientiously . . . to aim at the welfare of the whole population of the country. . . . Even if this Privy Council has no definite powers, it may nevertheless bring about the gradual composition of differences, and secure a united representation of the common interests of the country. (Munich, 1930)

On Right Wing Anti-Communists
The reactionary politicians have managed to instill suspicion of all intellectual efforts into the public by dangling before their eyes a danger from without. Having succeeded so far, they are now proceeding to suppress the freedom of teaching and to deprive of their positions all those who do not prove submissive. (New York, 1953)

On Human Rights

The existence and the validity of human rights are not written in the stars. The ideals concerning the conduct of men toward each other and the desirable structure of the community have been conceived and taught by enlightened individuals in the course of history. Those ideals and convictions . . . have been trampled upon. . . . A large part of history is therefore replete with the struggle for those human rights, an eternal struggle in which a final victory can never be won. But to tire in that struggle would mean the ruin of society. (Chicago, 1954)

The international significance of all the causes on which he spoke out enlarged his human figure to great proportions.

Einstein Looks for Order in a Chaotic World

Traveling on a Beam of Light

Albert Einstein was far more than a scientist who pondered space and time. Yet it is his soaring scientific mind for which we know him. He was born on March 14, 1879, in Ulm, Germany. About a year later, his father's business collapsed and the family moved to Munich with Einstein's uncle Jakob to set up a small electrical plant and engineering works. Einstein attended rigidly disciplined schools in Munich, where he developed his lifelong dislike for regimented education and mindless discipline, as seen in his numerous public criticisms of military authority and totalitarian states.

The stories about his poor scholastic performance in these early years are true. At school, he was bored, intimidated, shy, and withdrawn. Historians have even suggested that he suffered from a form of dyslexia, but that has not been confirmed. Einstein's brother once wrote that Albert's "teachers reported . . . that he was mentally slow, unsociable and adrift forever in his foolish dreams." Responding to the school's headmaster about what profession Albert might enter someday, Einstein's father said, "It doesn't matter . . . he'll never make a success of anything."

There were a number of influences in his early childhood that played a part in setting him on the path to relativity: a pocket compass, given to him at age five, which indicated to him that space was not empty, as he'd been taught; his love for the violin, which led him to an awareness of the mathematical structure of music; his uncle Jakob Einstein, who made mathematics interesting; and his uncle Casar Koch, who recognized and supported his interest in science. But it was Max Talmey, a young medical student at Munich University, and Einstein's tutor and mentor, who first sparked his latent brilliance by giving him advanced books on physics and mathematics. Talmey said, "His mathematical genius was so high that I could no longer follow. Thereafter philosophy was often a subject of our conversations. I recommended to him the reading of Kant." Later Einstein also read the philosophical works of David Hume, Ernst Mach, and others.

By the time he was twelve years old, he'd already developed a healthy skepticism for much of the material forced on him in his earlier years in the regimented German school system. He began to focus on the physical world and refine his thoughts about it. Einstein was expelled from school at age fifteen for refusing to submit to that regimentation. He was told by his instructor: "Your presence in the class is disruptive and affects the other students." He then proceeded to Milan, Italy, to rejoin his family, who had moved there from Germany because his father's business had failed once more. The next year, at age sixteen, while Einstein was riding his bicycle through the Italian countryside, he asked himself an innocent question that would lead to one of the greatest scientific discoveries in history—one that fundamentally altered our view of the universe: *What would the world look like if I were sitting on a beam of light, moving at the speed of light?*

Einstein wanted to continue his education in physics and math by obtaining admission to the renowned Polytechnic Academy in Zurich, Switzerland. However, in the spring of 1895 he was only sixteen years old and one had to be at least eighteen to take the entrance exam for this school. But that year he wrote a paper about electromagnetism that foreshadowed the theory of relativity and sent the paper to his uncle Casar. Through family friends, the paper made its way to the principal of the academy, eventually leading to Einstein's admission. His goal was to become a teacher of mathematical physics, so his subjects included differential and integral calculus, descriptive and analytical geometry, the geometry of numbers and the theory of the definite integral. In the final years of the nineteenth century, new theories about electromagnetism emerged and scientists recognized that Newton's physics could not explain certain phenomena. This new knowledge began to shake the foundations of accepted thought about the universe. During Einstein's four years at the academy, his thoughts were being formed simultaneously with Röntgen's discovery of x-rays, Thomson's discovery of the electron, and Ernest Rutherford's gradual formulation of the modern concept of the atom. Yet Newtonian physics had been accepted and had worked for so long and was so firmly entrenched that the conservative establishment of science was extremely reluctant to acknowledge any new concept that might reflect any shortcomings in that body of knowledge.

Einstein's ability to conceive new approaches without the shackles of accepted scientific dogma combined with the influences of his uncles and Max Talmey, the recent discoveries in the field of electromagnetism, and four years of education in physics and math at the Polytechnic Academy set the stage for one of the greatest intellectual advances in the history of science. In the meantime, however, Einstein continued to display his rebelliousness and independence at the academy. "You have one fault," a professor advised him, "one can't tell you anything." As a result, upon graduation he was passed over for a position as an assistant professor in the physics department at the academy, dashing his hopes to become a teacher of mathematical physics.

Einstein Redefines the Universe

An Unknown Clerk . . . the Equivalence of Mass and Energy

Einstein returned to Milan to live with his family again for several months, but soon returned to Zurich and became a Swiss citizen. With prospects of permanent employment greater for Einstein as a Swiss citizen in Switzerland than they'd been as a Jew in Germany, on June 16, 1902, Einstein was formally appointed as technical expert at the Swiss Patent Office in Berne, at a salary of 3,500 francs a year (the equivalent of about $3,000). He lived in a one-room apartment a few hundred yards from the office. His boyhood mentor, Max Talmey, visited Einstein in Berne and later wrote, "His environment betrayed a good deal of poverty. He lived in a small, poorly furnished room. . . . His hardships were aggravated through obstacles laid in his way by people who were jealous of him." Day after day, Einstein sat in the long narrow room of the patent office, along with his fellow technical officers, reading the endless patent applications for typewriters, cameras, engineering instruments, and the myriad other assorted devices for which hopeful inventors wished to claim legal protection. He soon mastered the undemanding routine, thereby freeing his creative genius to develop his ideas about physics in his spare hours away from the office.

To earn extra money, he placed an ad in the newspaper offering private tutoring in physics. This led to a following of bright young physics students two to three years his junior. Einstein socialized with his students and engaged in long intellectual discussions and debates that continued over the next several years. They often took trips together through the Swiss countryside and pursued their common passion for physics at Cafe Bollwerk, near the patent office, as Einstein refined his thoughts and improved his ability to express his theories in the most cogent manner possible. Some of these students later wrote that even in his early twenties not only was Einstein's genius apparent, he also had a commanding presence and impressive force of character.

In 1903, Einstein married Mileva Marić, a friend from his days in Zurich. Two years later, still an unknown twenty-six-year-old clerk in the Swiss Patent Office, he wrote an article, "On the Electro Dynamics of Moving Bodies," in which he answered the question he'd asked himself ten years earlier. Based on the mathematical calculations contained in the article, Einstein concluded that if light always moves with the same speed in free space, regardless of the motion of the source, the passage of time must be relative, not absolute. In other words, a person moving with light, at the speed of light, could click his or her own flashlight on, and the beam of the flashlight would leave that person and the flashlight *at the speed of light* relative to that person *and* relative to a *nonmoving observer*. The concepts and proof supporting this conclusion became known as the special theory of relativity (which will be examined in more detail in Chapter 8) and resulted in a fundamental change in how we view the universe. In a related paper, Einstein established the equivalence of mass and energy, represented by the famous equation $E = mc^2$, which was examined in Chapter 6. This was the most important conclusion that Einstein derived from the special theory of relativity.

In 1905, he published a total of five papers on entirely different subjects. Three of them were among the greatest in the history of physics. The paper concerning the quantum explanation of the photoelectric effect earned him the Nobel Prize sixteen years later. The third paper contained the special theory of relativity, adding space-time as the fourth dimension of the universe—9,000 words that compose one of the most remarkable scientific papers ever written.

Nevertheless, it took time for other physicists to appreciate the significance of Einstein's papers. In 1907, Einstein took the first step in what would become a long academic career when he became an instructor on the theoretical physics faculty at the University of Berne, while retaining his job at the patent office. There were only four students in his first class and *one* in the following term. As relativity slowly percolated through the scientific community and others began to appreciate what this new body of knowledge represented, Einstein was drawn out of his limited and isolated shell and was gradually embraced as a full member of the international scientific community. He began spending more and more time away from Berne, as he established relationships with Lorentz in Holland, Mach in Austria, Rutherford in England, Curie and Langevin in France, Planck in Berlin, and Sommerfeld in Born. By 1909 he was receiving numerous invitations to lecture at various scientific conferences and was soon regarded as a leader of the scientific elite of the early twentieth century that was focused on discovering the nature of the physical world.

On July 6, 1909, Einstein resigned from the patent office and moved back to Zurich to take up his first full-time academic position in the newly established chair of theoretical physics at the University of Zurich. Einstein's lectures at Zurich (on thermodynamics, the kinetic theory of heat, electromagnetism, and other topics in physics) became extremely popular, due to his humor, unusual presentations, patience, and accommodation to his students to make sure they understood.

Those next few years were dominated by his continued travel throughout Europe and by the refinement and consolidation of his scientific theories, including development of his general theory of relativity (which will also be discussed in Chapter 8). In March 1911, he accepted a position at the German University in Prague, Czechoslovakia, a few miles from the place where Tycho Brahe first employed the young Johannes Kepler. He noticed a growing hostility between the Czech and German citizens there, and unbeknown to him and to the world, this animosity foreshadowed an increasingly volatile political situation in Prague that would precipitate world war.

By this time the Einsteins had two children—Hans, born in 1903, and Edward, born in 1910. In March 1912, Einstein began what was to be a long friendship with Marie Curie (until her death in 1934) when he went to Paris to address the French Society of Physicists. In July of that year, their families took a joint vacation in the Alps. The Curies' eldest daughter, Irene, eventually took up the work of her mother and also established a long friendship with Einstein.

In April 1914, the family moved to Berlin, where Einstein began working with the generous support of Germany's premiere science association, the Prussian Academy of Sciences. Einstein expressed his gratitude in this letter:

> Gentlemen . . . I have to thank you most heartily for conferring the greatest benefit on me that anybody can confer on a man like myself. By electing me to your Academy you have freed me from the distractions and cares of a professional life and so made it possible for me to devote myself entirely to scientific studies.

These words and the mutual admiration between Einstein and the academy would soon be swallowed up in the darkness that engulfed the world as the Great War erupted. The changing face of the world and Germany's pivotal role in it immediately changed Einstein's life and led to his transformation from a quiet scientist to champion of pacifism.

Because of the war, Mileva and their two sons were not able to return to Berlin after their summer vacation in Switzerland in 1914. This involuntary separation eventually led to divorce. In 1915 Einstein began to speak out against German militarism and nationalism in general. He focused his life on two matters—the world war and finalizing his *general* theory of relativity. He first published that theory in 1916 in an article titled "The Foundation of the General Theory of Relativity." Gravitation, Einstein explained, is not a force, despite what Newton had thought. Instead, "it is a curved field in the space-time continuum." Einstein's excitement over concluding this work was revealed in a letter to his friend the German physicist Arnold Sommerfield (1868–1951) in November 1916, when he said, "This last month I have lived through the most exciting and most exacting period of my life, and it would be true to say that it has also been the most fruitful period." Other physicists recognized the combined theories of relativity as a revolution in thought, the likes of which had not been seen since Newton's *Principia*. In 2011, after circling Earth for fifteen months, NASA's satellite from its Gravity Probe B mission confirmed Einstein's theory of general relativity, that gravity arises when space and time are bent by mass.

Einstein Steps onto the World Stage

Hitler's Greatest Public Enemy

In 1919 Einstein married Elsa Lowenthal, a distant relative of his late father, and lived with her and her two daughters in Berlin. In that same year, the Royal Society of London announced it had photographed the solar eclipse and completed calculations verifying the predictions that Einstein had made in his general theory of relativity concerning the deflection of starlight. Although Einstein had already become well known in scientific circles throughout the world, true international fame came to him as a result of the Royal Society's announcement. His public life changed dramatically. He became the symbol of science and the master of twentieth-century intellect. For the next thirty-five years, he became outspoken on the long list of social causes mentioned earlier, primarily centered around his promotion of international pacifism. Through the 1920s, a period of unrealistic ideals, Einstein embodied idealism and served on a number of organizations that promoted his philosophies, such as the newly formed League of Nations' International Committee of Intellectual Cooperation, along with Marie Curie and other physicists.

By 1929, Einstein had established strong ties with physicists all over the world, including the United States. With the resurgence of German militarism, accompanied by an unprecedented and most sinister level of anti-Semitism, Einstein began to think of not only leaving Germany but fully detaching himself from the entire European continent. By mid-1932, when it was obvious to Einstein that the Nazis would soon gain power in Germany, Einstein and his wife were already making plans to move to the United

States. While on a lecture tour there in early 1933, Einstein's fate was sealed when on January 30, 1933, Adolph Hitler came to power as chancellor of Germany. Einstein then renounced his German citizenship, condemned the country of his birth, and urged all of Europe to take up arms against the Nazis and prevent them from precipitating a war. His predictions in these early years of Hitler's reign were ignored. In retaliation for Einstein's public statements, Nazi storm troopers ransacked his summer house outside of Berlin. Einstein's relationship with the Prussian Academy of Sciences, which had embraced him in 1914, abruptly ended when Einstein publicly withdrew and the academy issued its Declaration of April 1, 1933, in which it vowed allegiance to Germany and stated that it had "no reason to regret Einstein's withdrawal."

In October 1933 he accepted a full-time position at the school of mathematics in the newly founded Institute for Advanced Study at Princeton University, in New Jersey, where he would remain for the following twenty-two years. As if the special and general theories of relativity had not been great enough accomplishments, Einstein pursued the question of whether it would be possible to unite those and other theories into what would be called the "unified field theory." This scientific pursuit would dominate his remaining years but would never bear fruit.

His scientific and social pursuits were not nearly as influential or significant after the 1920s as they'd been up to that time. In addition, his son Edward suffered a mental breakdown and blamed his father for deserting him after the 1914 divorce from Mileva and for being the cause of his troubled life. After this crushing development and after the death of his wife, Elsa, in 1936, Einstein became increasingly solemn and discouraged by his own life, including his inability to slow the threat of war that he knew would be disastrous to so many innocent people. As seen in Chapter 6, his quiet work at Princeton was interrupted in 1939 when he agreed to sign the letter to Roosevelt, which led to the Manhattan Project. Otherwise, his travels lessened through the 1940s and he maintained a routine existence centered on his theoretical research, but he didn't make any further spectacular findings in physics.

Measured by his lasting contribution to our understanding of the universe, Albert Einstein will always remain a great figure in science. But to fully appreciate him as a human being, one must also understand his other passions—peace, the human condition, truth, and the difficult task of creating political institutions that maintain freedom in society. Einstein reprimanded the failed governments and the oppressive policies that determined millions of people's fates during the first half of the twentieth century and became the conscience of the world. But those crusades failed. Ironically, his only effort that resulted in the establishment of a clear government policy was the fateful letter to Roosevelt encouraging the establishment of the research that led to the Manhattan Project, forever linking Einstein's name to the production of the atom bombs that killed over 200,000 people and destroyed Hiroshima and Nagasaki in August 1945. Einstein expressed regret about this in many different ways at different times. For example, he said to Linus Pauling, "I made one great mistake in my life—when I signed the letter to President Roosevelt recommending that atom bombs be made." He also wrote, "The physicists who participated in forging the most formidable and dangerous weapon of all times are harassed by a feeling of responsibility, not to say guilt." He saw himself and those other physicists as collaborators in the invention of the worst death machine in history. Yet, in the context of the time when he made the decision to sign the letter, it was a logical decision, one that he sometimes defended:

> We helped in creating the new weapon in order to prevent the enemies of mankind from achieving it ahead of us, which, given the mentality of the Nazis, would have meant inconceivable destruction and the enslavement of the rest of the world. We delivered this weapon into the hand of the American and British people as trustees of the whole of mankind, as fighters for peace and liberty. But so far we fail to see any guaranty of peace. We do not see any guarantee of the freedoms that were promised. . . . The war is won, but the peace is not.

Thus, based on the possibility that the Germans would develop the bomb first, he felt the Manhattan Project was justified. But when the United States later discovered that the Germans were not as far along as they feared, and when the war in Europe ended in May 1945, he was opposed to using the weapon. "I have always condemned the use of the atomic bomb against Japan," he said. Einstein's dedication to pacifism, which became an even greater passion after the bombs were dropped on Hiroshima and Nagasaki, centered on the belief there should be one world government under a constitution to be jointly drafted by the United States, Britain, and Russia. "I advocate world government," wrote Einstein, "because I am convinced there is no other possible way of eliminating the most terrible danger in which man has ever found himself."

Einstein's position on this point was so naive and unrealistic that one must question his knowledge and understanding of the powerful social, cultural, and political forces that bind groups into national and regional entities. These forces have roots that go back centuries or even thousands of years and rear their ugly heads regularly around the world—places like Bosnia, Chechnya, the Middle East, Somalia, Rwanda, Haiti, and Northern Ireland. They cannot be subdued by a vague and untested prediction that by nations universally relinquishing sovereignty the greatest good for the greatest number of people will result.

Although Einstein's social and political views never had a significant impact on those who controlled governments, he remained a respected figure through most of his life. However, after World War II, his pacifist and one-world-government views, in a nation increasingly opposed to and distrustful of the Soviet Union, led to a broad distrust of Einstein himself. He suffered a final though gentle fall from grace after his address to a television audience in 1950:

> And now the public is advised that the production of the hydrogen bomb is the new goal. . . . An accelerated development towards this end has been solemnly proclaimed by the President. If these efforts should prove successful, radioactive poisoning of the atmosphere, and, hence, annihilation of all life on earth, will have been brought within the range of what is technically possible.

In his waning years following this speech, he persisted in his call for pacifism, yet was viewed increasingly out of context, out of his element, and as a lone, eccentric scientist out of touch with social and political reality—a view largely supported by his description of himself:

> My passionate interest in social justice and social responsibility has always stood in curious contrast to a marked lack of desire for direct

association with men and women. . . . I have never belonged whole-heartedly to any country or state, to my circle of friends, or even to my own family. These ties have always been accompanied by a vague aloof-ness, and the wish to withdraw into myself increases with the years. Such isolation is sometimes bitter, but I do not regret being cut off from the understanding and sympathy of other men. I lose something by it, to be sure, but I am compensated for it in being rendered independent of the customs, opinions, and prejudices of others. . . . It is strange to be known so universally and yet to be so lonely.

Albert Einstein had no illusions about whether his words could sway nations. But he was morally compelled to speak and write those words. He wanted "simply to serve with my feeble capacity truth and justice at the risk of pleasing no one." Through his devotion to the activities and state of humanity, and as a person profoundly concerned with science as a part of our culture, Einstein—the German who hated Germany, the Zionist who tried to make peace with the Arabs, the pacifist who had a hand in the birth of nuclear weapons, and the introverted loner who became a spokesman on the world stage—demonstrated better than anyone how dependent science is on a free society and how fleeting that freedom can be.

Chapter Eight
THE FOURTH DIMENSION

Sometimes I see
clearest of all
when I'm alone and remember
afternoon sunlight at schoolday's end
the world through the eyes of a child.

Then I can see
a world yet to be
here and now in my grasp,
like Saturday mornings when I wake and see
the world through the eyes of a child.

When I grow old
in Winter and twilight
I will remember Spring
and wisdom I learned
from that sometime view of
the world through the eyes of a child.
　　　　　—John Kuzma, "The World Through the Eyes of a Child" (1996)

Everyone knows that Einstein was a scientific genius responsible for doing something extraordinary, but few people know what it was. This chapter will attempt to explain relativity. To *fully* understand it, one needs knowledge of advanced physics, but like the six other great scientific discoveries covered in this book, relativity can be grasped conceptually and without equations. Yet, relativity does require a higher level of abstract thinking than is necessary to understand the other discoveries. We don't experience relativity like we do gravity, for example, and we can't *see* relativity like we do the galaxies or the human cell. Einstein himself said that the advantage of most theories is their clarity, while the advantage of proving and understanding relativity is its "logical perfection." The principle of relativity is the one most at odds with common sense, so it requires looking at the world unfettered by a lifetime of experiences. It requires seeing the world through the eyes of a child.

Newtonian Physics Falls Short

Questions about Mercury's Orbit

We will first examine Einstein's special theory of relativity. At the end of this chapter we'll briefly look at his general theory of relativity. *The central significance of the special theory of relativity is that the passage of time is not absolute.* "Absolute, true and mathematical time," Newton wrote in the *Principia*, "of itself and from its own nature, flows equably, without relation to anything external." Remarkably, numerous experiments testing the special theory prove that Newton's statement is not true. The rate at which time passes is different for one person "at rest" than it is for another who is *moving* at great speed in relation to the person at rest, even though for both of those people *their* time is the real, actual, and normal time. With the discovery of relativity, we realized space and time are not separate. They are intertwined as "*space-time.*" There was nothing wrong with Newton's physics, but it had to be extended to explain the relationship between velocity and space-time. Special relativity didn't overthrow Newtonian physics, it only demonstrated that Newton's principles were not valid in extreme conditions. When Newton referred to distance, he meant distance at a given time. No one, including Newton, thought there could be any ambiguity about *time*.

The laws and principles discovered by Newton and revealed in the *Principia* in 1687 had provided all the answers necessary for physics to operate successfully for over 200 years. Galileo and Newton studied relativity as that term was used in their time. Some of the ideas they developed were an elaboration of concepts known since ancient Greece, while others were a correction of ancient beliefs. In all cases, Galileo's and Newton's laws on this subject were universally accepted before Einstein's theory and are known as "Galilean relativity." Here are two examples of Galilean relativity:

1. If a person is walking up the aisle of a train, in the same direction as the train, his or her speed *relative* to the ground is the sum of the train's speed and the walking speed. For example, if the train is moving at seventy miles an hour, and the person is walking at four miles an hour, the person's speed *relative to the Earth* is seventy-four miles an hour.

2. If a person on a moving train drops an object from the ceiling to the floor, it will drop straight down relative to the train and to the person on the train. However, relative to Earth and as seen by an observer who is standing ("at rest") outside the train, it will appear to drop in a motion that looks like a *curve* because the vertical velocity of any falling object increases with time. This parabolic curve is shown in Figure 8-1.

These examples of Galilean relativity show that some type of *frame of reference* (such as the train, the Earth, another planet, or other heavenly body) is needed to describe the movement of an object.

FIGURE 8-1
Galilean Relativity—Object Dropped in Moving Train

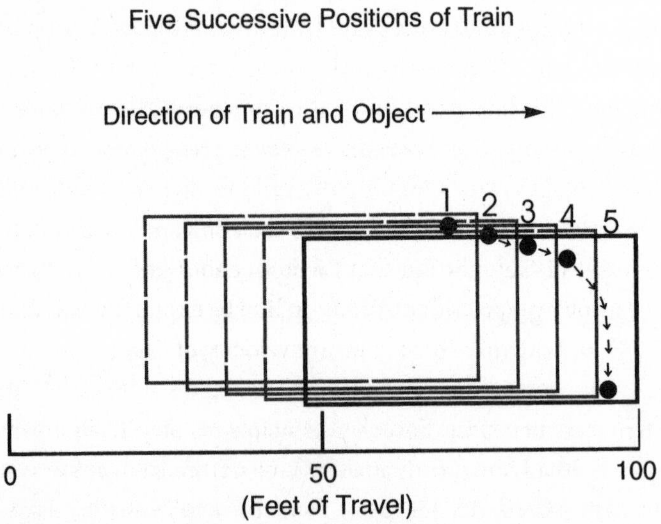

Five Successive Positions of Train

Direction of Train and Object ⟶

0   50   100
(Feet of Travel)

Toward the end of the 1800s facts were slowly being uncovered that seemed to conflict with or were not addressed by the prevailing laws of physics, including Galilean relativity. For example:

The speed of light did not behave like speeds of material objects.

The heat of molecular gases differed from values calculated using Newtonian theory.

Radioactivity showed that matter was subject to unpredictable eruptions of instability.

Mercury's orbit was at variance with Newton's laws of motion.

There was no explanation in classical physics for the existence of light and other radiation emitted by atoms.

For the first time since Newton, scientists began to question the validity of the classical framework in which they had put so much faith. Their instruments were becoming more refined and sophisticated, thus enabling them to identify these ever so slight discrepancies. One discrepancy in particular—the mystery that surrounded the first point in the list, regarding the speed of light—became explainable when Albert Einstein conceived his special theory of relativity in 1905. Again, the central significance and the greatness of the special theory is that *the passage of time is not absolute.* The foundation of the theory is based simply on Einstein's combination of two principles: Galilean relativity plus a principle that he discovered called the constant velocity of light. This principle holds that light always has the same velocity (186,000 miles per second), *regardless of the motion of the observer or the source of the light.*

Einstein's discovery of the principle of the constant velocity of light was the direct result of the question that Einstein had asked himself at age sixteen. In his autobiographical notes later in life, he explained why he became so certain of the principle of the constant velocity of light:

After ten years of reflection such a principle resulted from a paradox upon which I had already hit at the age of sixteen: If I pursue a beam of light with the velocity [of light], I should observe such a beam as a spatially oscillatory electromagnetic field *at rest*. However, there seems to be no such thing. . . . From the very beginning it appeared to me intuitively clear that . . . everything would have to happen according to the same laws as for an observer who, relative to the earth, was at rest. For how, otherwise, should the first observer . . . be able to determine that he is in a state of fast uniform motion?

The Scottish physicist James Clerk Maxwell (1831–1879) had published his two famous papers in 1861 and 1865 on the existence of an invisible electromagnetic field that included a broad range of *waves* (x-rays, radio waves, light waves, etc.). In other words, Maxwell had shown that light waves are simply the *visible* portion of this spectrum and that all these waves travel at 186,000 miles a second. Einstein theorized that the speed of these electromagnetic waves (including light) must be a constant and not a relative quantity. The speed of the observer shouldn't make a difference in the speed of such waves. Therefore, either Maxwell's equations were incorrect or Newtonian mechanics was incorrect. The need to resolve this inherent conflict led him to the special theory, which asserts that Maxwell's equations are correct and Newtonian physics is not adequate to explain space and time when an object or observer approaches the speed of light.

The Michelson-Morley Experiment Is Resurrected to Support Relativity

Vibrations in the Ether

The constant velocity of light was actually demonstrated twenty years before Einstein stated it as a principle. From 1881 to 1887, the German-born American physicist Albert A. Michelson (1852–1931) conducted a series of experiments using light, as described later. This work culminated in a final and now famous experiment in 1887 by Michelson and the American chemist Edward W. Morley (1838–1923), which became known as the Michelson-Morley experiment. Although these experiments were aimed at proving a different theory and were not correctly interpreted as applying to the speed of light until Einstein came forth with the special theory of relativity, they later provided the best data available to support Einstein's postulate on the constant velocity of light, which is critical to understanding relativity.

Even after Maxwell's discovery of electromagnetic radiation, most physicists also believed that space was filled with an invisible weightless substance that actually oscillated, in wave form, in order for light to propagate, similar to sound resulting from the vibration of air. The supposed substance was called "luminiferous ether" (or just ether). Despite Maxwell showing that light itself is an electromagnetic phenomenon—that is, waves and not particles—most physicists continued to believe in some type of ether to "support" or conduct the light waves.

The early experiments by Michelson and the Michelson-Morley experiment of 1887 were intended to confirm the existence of the ether. They tried to prove this by measuring the difference in the time it takes light to travel in the direction of the Earth's motion in its orbit around the sun compared to ninety degrees (or transversely) to that direction. In the framework of Galilean relativity before it was known that the speed of light is always constant, one would expect there to be a difference because, like the person walking up the aisle of a train or the object being dropped on a train, a beam of light directed *toward* the Earth's direction of motion around the sun would be expected to travel at the speed of light *minus* that orbital movement of about 67,000 miles an hour, like a boat moving upstream in a river. It could then be compared to the speed of light traveling in other directions, and from those data the speed of the Earth relative to the ether—the object of the experiments—could be calculated. In 1887, Michelson and Morley (both working in Cleveland at the time) repeated the earlier experiments by Michelson and produced the same result—namely, *no detection of the Earth moving through the ether, no detectable change in the speed of light.*

If one forgets about the ether theory, as Michelson, Morley, and other scientists should have done, the experiments showed that the speed of light was the same in all directions at all times. That is, light doesn't change its velocity depending on whether it's traveling "upstream" or "downstream" through the nonexistent ether. With modern technology, the principle of the constant velocity of light has been confirmed beyond doubt.

In 1907, Michelson became the first American ever to receive the Nobel Prize in the sciences, for his spectroscopic and metrological (science of measurement) discoveries. Even though Michelson's work is now regarded as very significant in the history of physics, he felt that the experiments were a total failure because he never achieved his objective of measuring the speed of the Earth relative to the ether.

Einstein Determines That If Light Speed Is Absolute, Time Is Not

Moving Frames of Reference

Given the principle of the constant velocity of light, we'll now return to special relativity, remembering its central significance: *the passage of time is not absolute.* The special theory pertains only to the relationship between two objects, where one is moving toward or away from another at great speed— that is the movement of one relative to another—thus the name for the theory. Obviously, there would be no *relativity* in a universe with only *one* object in it because there's nothing to be relative *to.* But in our universe, if one object or observer is at rest *in relation* to another object or observer that is moving at great speed, the movement of that second object or observer *in relation to the first* results in a difference in the passage of time for each observer, as measured by the other. If observers A and B are at rest in relation to each other, but C is moving at great speed relative to A and B, then the passage of time for C will be measured by A and B to be slower.

FIGURE 8-2
A–B Frame of Reference Relative to C

C is moving away from the A-B frame of reference

A is at rest relative to B   B is at rest relative to A

The special theory of relativity began in Einstein's mind as a concept. He then created the mathematical proof to support it. Numerous experiments in the twentieth century have provided the supporting empirical evidence. We will not delve into any of the math or the experiments, but we can gain a clearer conceptual understanding by referring to Figure 8-2 and considering A, B, and C as *objects or observers*, with A and B in one frame of reference—that is, a physical area in space or a body where A and B are stable and not moving *in relation to each other*—while object or observer C is in another frame of reference.

There are actually three ways to look at the A–B–C relationship when the space between A–B is increasing relative to C: (1) C is moving away from A–B, (2) A and B are moving away from C, or (3) A–B and C are moving away from *each other*. We will examine this as if option 1 were occurring— that is, A and B are "at rest" and C is moving, as shown in Figure 8-2. We'll use this as a model in this chapter so we have a consistent framework and viewpoint, even though options 2 and 3 are equally valid.

Light moving between A and B in one frame of reference covers a certain distance in a certain measured period of time.

However, what if C is rapidly moving away from the A–B reference frame, as suggested in Figure 8-2? Given that the speed of light does not change, regardless of the movement of the body from which the light originates, how do we reconcile the speed of a light beam flashed from A to B (that is, *within* one frame of reference) with what you would expect to be a *slower* speed of light flashed from the A–B reference frame *toward* C, as

FIGURE 8-3

Light within the A–B Frame of Reference

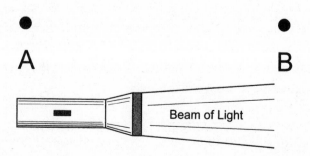

A   B

Beam of Light

FIGURE 8-4

Light Traveling from A–B to C

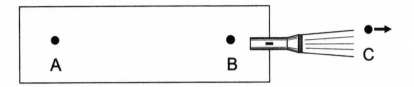

measured in frame C? That is, you would expect it to be slower because if Galilean relativity were applied to the two rapidly separating frames of reference, a "net" speed of light would result by subtracting the speed of the separation from the speed of light.

But we are given the principle of the constant velocity of light, and it must *somehow* be reconcilable with the relative movements of the two frames of reference. The special theory of relativity succeeded in this reconciliation by proving that time is "personal" to observers in separate inertial systems. That is, because light's actual speed over a distance is constant, the time element must *not* be constant. It must depend on *the observer's relationship to the moving object.* We are all in the same frame of reference on Earth and traveling slowly—therefore, time is constant and does not change. But if we were in reference frame A–B, and C was moving toward us at the speed of light, and if C flashed a beam of light toward A–B, light would *not* be traveling toward A–B at *twice* its normal speed (that is, 372,000 miles a second). Travel speed and light speed can't be *added* together because the velocity of light is constant.

The theory can be further understood by answering and examining these three questions:

1. How fast do you have to be going before your speed has an impact on how fast time passes relative to someone at rest? If there is one clock in the A–B frame of reference and another clock in frame C, how fast does C have to go before there is a significant difference in the rates of the two clocks as measured from either frame?

2. How can a person visualize and gain a clear understanding of this new perception of time?

3. What happens to matter physically when it's moving at these great speeds?

Once you understand the answers to these three questions, you will understand special relativity on a conceptual basis.

QUESTION NUMBER 1
What Is the Relationship between Speed and Time?

The Relativistic Factor

Remarkably, Einstein's theory would give a different result of the speed of the person walking down the aisle of the train than would Galilean relativity, though the tiny difference would be extremely difficult to measure. But if we are dealing with observers and objects under observation that are moving *relative to each other* at velocities more than about half the speed of light,

FIGURE 8-5
Correlation Between Speed and Time Dilation

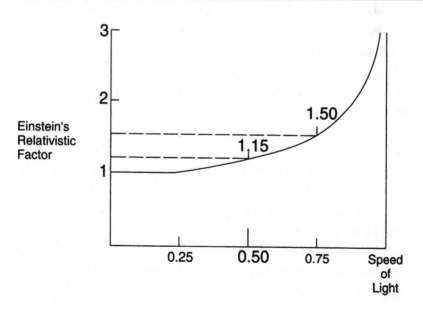

Einstein's theory is the only correct way to determine the result. This is shown in Figure 8-5.

The commonly observed phenomena in our daily lives are adequately described by Galilean relativity and Newtonian physics. However, the graph in Figure 8-5 shows that once any matter approaches the velocity of half the speed of light, the effect on time becomes increasingly dramatic because there is an increasing and much more noticeable effect of speed on time "intervals"—that is, the time *between* events, the very rate at which time passes. So, for example, the time dilation at half the speed of light is noted in Figure 8-5 and shows that if an object or observer (such as C) is moving at that speed, time intervals in C's frame of reference will be 15 percent shorter than the time intervals between the same events as measured in frame A–B, which is the frame of reference at rest. One hour for a person moving at half the speed of light is equivalent to one hour and nine minutes for a person on Earth, because 115 percent of sixty minutes is sixtynine minutes. This ratio between the two time periods is calculated by multiplying C's time by Einstein's "relativistic factor" (which is derived by an equation he developed). A speeding space traveler's life will actually be extended relative to an Earth-bound person, even though the passage of time for the person traveling at half light speed and everyone with him or her will *feel* no different because that will be their *actual* time. It is not a *perceived* change in time. It is their *actual* time. It has been proven that moving clocks run slower.

The idea that time can actually change and is a function of velocity is totally contrary to our everyday experience. Time seems so firm, so constant, and so "un-relative" that any tinkering with it is difficult to accept. Under Newtonian physics, *mass, length* of any physical object, and *time* were the sole fundamental quantities needed to describe mechanical movement of such objects. Einstein added his "relativistic factor" and showed that as such movement approaches the speed of light, the time that *actually* (not theoretically) passes in the moving frame will be less, relative to the actual time that passes in a frame of reference at rest.

The purpose of the relativistic factor is to *adjust* or *correct* for the two time frames. Without such an adjustment, we would be left with the irreconcilable results that do not account for the principle of the constant velocity of light. Indeed, Einstein's principle would be more accurately called "abso-

FIGURE 8-6

Two-Minute Intervals Between Flashes

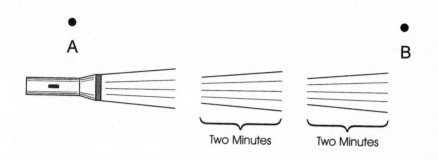

lutism" or "the special theory of *non*-relativity" because he was seeking a *unifying* principle and the *removal* of the relative nature of the fundamental laws of Newtonian physics. He did not name his principle the special theory of relativity—it was a term invented by others.

One more example will further illustrate the relationship between speed and time and that the passage of time is not absolute. Imagine that in the A–B reference frame, with A and B at a great but constant distance apart from each other, A aims his flashlight at B and turns it on for an instant once every two minutes, sending a *white* beam of light to B. Thus B sees the beginning of each successive flash every two minutes.

Because A and B are in a constant position relative to each other, the first instant of each flash takes the same amount of time (two minutes) to reach B. We don't know the time it takes for the *first* flash of light *to travel from A to B*. It could be a minute or ten years depending on the distance between A and B, but that doesn't matter in this situation because all we are concerned with is the two-minute *interval between* the *beginning* of one flash to the beginning of the next.

Now add to this example the assumptions that person C is traveling *away from* A toward B at a constant speed, and while C is moving she's watching A's flashes (Figure 8-7). Of course, each flash reaches C before it reaches B because she's closer to A, and—even at a very slow speed—because C is moving *away* from A, it takes *more* than two minutes *between* flashes.

In other words, if C had stopped at the exact location where she was when the first flash reached her, the second flash would have come two

FIGURE 8-7

C Traveling from A to B

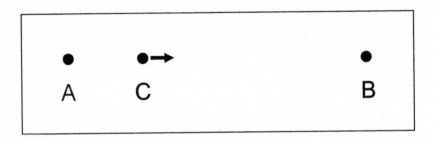

minutes later—the same *interval* as A's actual flash and B's receipt of it. But because C is moving away from A and each succeeding flash has to travel farther than the previous one to reach C, the second flash would take more than two minutes to get there. This would happen without regard to any relativity principles—Galilean or Einstein's. The interesting phenomenon is not what happens between A and C, but rather the effect of special relativity between C and B, as described in the following paragraph and as shown in Figure 8-8.

Now add these assumptions to this example: C's speed is 75 percent of the speed of light, and C is carrying a light source of her own and is sending a second beam of light, let's say *yellow*, to B beginning at the precise moment that A's white beam gets to C (Figure 8-8). Although C is moving, her yellow flash would still leave at light speed and travel to B at light speed. Thus will C's yellow beam reach B *before* A's white beam? The answer is no!

As discussed earlier, it would be impossible for the two beams to reach B at different times. The beams arrive *at the same time* because they are leaving C at the same time and are traveling together from C to B *at the same time*, even though they emanate from sources that are not in the same frame of reference—white beam from A and yellow beam from C. But C is moving very fast!

Applying Einstein's relativistic factor shown in Figure 8-5, at 75 percent of the speed of light, the same two-minute interval for A and B would increase 50 percent for C. Thus it would be a *three-minute* interval for C. How can the two beams reach B together when the white beams or flashes (A to B) reach B every *two* minutes in the A–B frame of reference, while the

FIGURE 8-8
C Traveling from A to B and Sending Beam to B

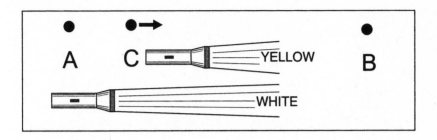

yellow beams or flashes (C to B) reach B every *three* minutes? The only answer is that "time" for A and B is not the same "time" for C. *Because the white and yellow beams of light require two different time periods in order to cover the same distance and in order to be explained to two separate observers, then the time component must not be constant.*

As long as A and B are in a constant (unmoving or inertial) relationship *to each other* their time intervals will match. But their perception of elapsed time will never be the same as that of an observer moving (at great speed) toward or away from them. This is the only possible explanation for the fact that B sees A's white beam arriving at B the same time as C's yellow beam, even though C's yellow beam has the "advantage" of C's *additional* speed toward B. From B's perspective as the receiver of both beams, they look exactly the same—at the instant the white beam reaches and then "leaves" C, the yellow beam travels with it. *They must reach B together every two minutes because the speed of light is constant.* Yet, from C's perspective, the beams are leaving her every *three* minutes. The time *interval* has increased 50 percent in speeding traveler C's reference frame. The rate of the passage of time for C is slower than for A and B.

Returning to Galilean relativity for a moment, if it were a white *ball* being thrown from A to B instead of a beam of light, with C launching her own yellow ball toward B at the moment the white ball gets to her, of course, C's speed would be used in calculating how fast she would have to throw the yellow ball to keep it astride the white ball as it goes the remaining distance. So, for example, if C was going twenty miles per hour and the white ball was going fifty miles per hour, C would have to throw the yellow ball thirty miles per hour, resulting in both balls moving forward together at fifty miles per hour for the remaining distance to B. But remember, based on Einstein's theory of special relativity, we have removed the speed variable from the equation (speed of light is constant) and removed the ability *to combine the speed of the source* from which the light originates.

QUESTION NUMBER 2
How Can This Concept of Time Be Visualized?

Banking Time in Canadian and American Dollars

To conceptualize time dilation we can compare the two frames of reference to the exchange rate of two different countries. As all foreign travelers experience, on any given day the currency of one nation is equivalent to a fixed amount of currency of each other nation in the world. Imagine that all Earth-bound people at rest (such as A and B) are American or British. Then assume that C is a person in space traveling at half the speed of light and that all such space travelers are Canadian. Further assume that all commodities for sale (including the precious commodity of time) are priced the same in both the United States and Canada. In other words, one hour of time can be purchased by Americans for the same number of U.S. dollars as Canadians would need in Canadian dollars. In this hypothetical situation, assume that the following exchange rates between the countries apply:

- The American (U.S.) dollar is equivalent to $0.87 in Canadian money. In other words, it would take only $0.87 for C, a Canadian (the space traveler) to purchase an American dollar or $1.00 worth of American goods.

- Reversing the ratio, A and B who are American or British (at rest on Earth) would have to pay $1.15 (American) to purchase $1.00 Canadian or to purchase $1.00 worth of Canadian goods. In other words, an American would have to use more dollars than a Canadian to purchase the same item.

FIGURE 8-9

Foreign Exchange Rate Compared to Time Dilation

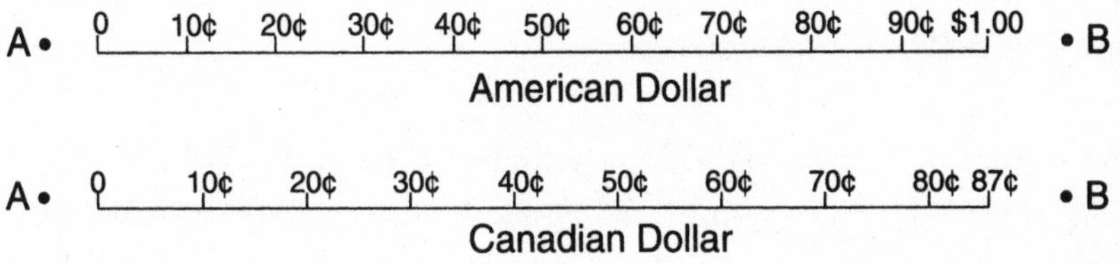

A•
| 0 | 10¢ | 20¢ | 30¢ | 40¢ | 50¢ | 60¢ | 70¢ | 80¢ | 90¢ $1.00 |
•B

American Dollar

A•
| 0 | 10¢ | 20¢ | 30¢ | 40¢ | 50¢ | 60¢ | 70¢ | 80¢ 87¢ |
•B

Canadian Dollar

This comparison is illustrated in Figure 8-9.

If we now substitute the concept of time in place of the dollars, we see that C, the space traveler, and other Canadians are able to buy more "time" than A and B, the Americans and British on Earth—that is, the Canadians (the space travelers) have the more valuable currency. They have greater purchasing power. Converting the 15 percent greater amount of money into 15 percent greater amount of time (again, 115 percent of sixty minutes is sixty-nine minutes), it takes C less time (fewer dollars) to traverse the distance from A to B because her time (her dollar) is worth more in relation to the Americans and British. Thus, while it would take an American or British individual 100 years to traverse the distance between A and B, it would take a Canadian (C) only about 87 years to go that same distance (as measured by A or B), because her actual time has slowed relative to the Americans' and British residents' actual time. Just like C would need only $0.87 to purchase the $1.00 item, she would need less time (measured by anyone in the A–B frame of reference) to traverse the same distance.

Thus, in effect, the Canadian space traveler (C) is banking "spare time" just like she would be banking spare dollars in connection with the exchange rate between the two frames of reference (Canada and the United States). By banking this time, C's life as a space voyager would be extended relative to the life spans of A and B who remained on Earth, a popular concept experienced by numerous light-speed travelers in science-fiction adventures.

What Happens to Matter Physically at These Great Speeds?

Watching Orbiting Electrons

The answer to question 3 is nothing. However, even though nothing *happens* to such matter, if we examine the *physical characteristics* of matter itself when it moves at great speed, the concept of relativity becomes even clearer. We'll relate the theory to the physical aspects of matter in three different ways:

A. The correlation between all movement and time dilation, including the actual movement of subatomic particles.

B. The "foreshortening" (along the direction of motion) of matter traveling at great speed, as viewed by an observer in a different frame of reference.

C. Einstein's equation $E = mc^2$, which shows that the mass of an object is equivalent to energy.

A. Movement of Subatomic Particles

Imagine any object or particle of matter on the Canadian's spaceship, such as a hydrogen atom with its electron orbiting the nucleus at a given speed, and further assume an identical hydrogen atom in the frame of reference of the American or British person stuck on Earth. If each person had equipment that could simultaneously observe or otherwise record and measure the number of orbits that each of the electrons made around those two hydrogen nuclei during a period of time, *the actual specific number of orbits would be different for the two atoms.* The observers in the A–B frame of reference at rest would conclude that the electron revolving around the atom on the spaceship is *slower* relative to the electron in the atom on Earth—in the same ratio that could be calculated with the relativistic factor used under question 1 (correlation between speed and time dilation) and consistent with the time dilation concept discussed under question 2 (Canadian and American dollars). Both people could "see" this relative difference because their equipment would agree on the measurements. Of course, Einstein did not envision this hypothetical, because the structure of the atom was not known in 1905. However, it is an inescapable conclusion.

B. Foreshortening

The second point about the physical effect of moving at great speed is the "foreshortening" of matter in the direction of motion. As we explored in Part Two, matter is composed of atoms whose structure is kept in shape by electrical forces. In the modern concept of the atom's structure, the electrons are considered to vibrate in *waves* around the nucleus, rather than as discrete *particles*. Each atom has a certain radius between the electron waves or cloud and the nucleus, which accounts for the specific distance between atoms. If the distance between them changes, the overall length and shape of the object changes.

Thus the length of a one-foot ruler is determined by such average radii of the atoms of which it's composed. To an observer measuring that ruler as it goes by at a speed approaching the speed of light (but not to a person traveling with the ruler), the distance between those atoms and therefore the ruler itself will appear *foreshortened in the direction of its motion*. Thus, remarkably, even the shape (including the length) of measuring devices depends on their state of motion. Even the distance *measured* between points A and B in the example involving the white and yellow beams of light (Figure 8-8) would depend on whether that distance was being measured by A or B on the one hand or by C, the person traveling at great speed relative to them.

Observers in relative motion assign different numbers to length and time intervals between a pair of events, rather than finding these numbers to be absolute. Mass, length, and time intervals are all affected when measured from a moving frame of reference. Einstein's relativistic factor depends on the speed of an object—and mass, length, and time depend on the relativistic factor. According to Newton, of course, there was no "correction," because mass, length, and time were believed to be independent of speed. The entire geometry of space and the rate at which time passes are both determined by the speed at which matter moves through space. Again, the rate of "time" passage is not an absolute phenomenon—instead, it is the rate of physical change in the universe, as compared between two frames of reference in constant speed toward or away from each other. Such change includes the movement of hands on a clock, the flow of a stream, or any other physical movements that could be used to define the concept of "time." This further demonstrates that relativity involves a *physical* reality, not a *mental* perception. It applies without regard to our minds or senses.

C. Mass Converted to Energy

In the course of developing the special theory, Einstein discovered the most famous equation in science: $E = mc^2$, or energy (E) equals mass (m) times the speed of light (c) squared. (To square a number, just multiply that number by itself. Thus c^2 in this equation equals the huge number produced by multiplying 186,000 by 186,000, which is close to 36 billion.) While working on the special theory, Einstein realized that the inertia of a system (that is, one not rotating or accelerating relative to another system) *depends on its energy content.* Thus it struck him that inert mass is simply latent energy. This is the point expressed in the famous equation, which combines the principle of conservation of mass with the principle of the conservation of energy, as examined in Chapter 6 in connection with the development of atomic energy and the bomb. Never before realized or recognized, mass and energy are simply and literally two sides of the same equation. All mass carries energy and all energy contributes to mass. Energy in the form of light and other electromagnetic radiation (called photons) is not matter because it's not made of atoms. But even photons behave as if they had mass. The energy associated with even a small amount of mass is enormous, as we realized through the release of the enormous power in the atom's nucleus.

The atom's *mass* (thus matter in general) increases by the relativistic factor as its velocity approaches the speed of light. The objects moving at great speed have substantial energy of motion (kinetic energy), which gives them this additional mass. Again, energy is convertible into matter. While mass was constant in Newtonian physics, it becomes velocity dependent in relativity, requiring that mass is multiplied by the relativistic factor to obtain an accurate value of mass moving at great speed. In other words, the relativistic factor *equalizes* the two masses and "corrects" the relationship between the objects being compared. This *physical* difference between the matter at rest versus the matter moving at great speed is part of $E = mc^2$ and part of the very nature of matter. Referring back to the Canadian and American dollars in question 2, the "mass" of the American dollar is devalued in comparison to the increased "mass" of the Canadian dollar.

The General Theory Explains Gravity

Gravitational Waves Bending Light

There is a misconception about the general theory of relativity—namely, that few people in the world understand it because it is so difficult to comprehend. Though this might have had much truth to it when Einstein first introduced the general theory in 1916, it is no longer the case. Physics professors around the world teach the theory, scientific articles on relativity number in the thousands, and there are numerous books for laymen devoted to various aspects of the subject. However, the general theory of relativity does require familiarity with certain principles of mathematical physics and is more difficult to grasp conceptually than the special theory. Therefore, we will only briefly summarize it here.

Over the course of the ten years after his 1905 paper on special relativity, Einstein expanded that theory into the general theory of relativity. He explained:

> The theory of relativity resembles a building consisting of two separate stories, the special theory and the general theory. The special theory, on which the general theory rests, applies to all physical phenomena with the exception of gravitation; the general theory provides the law of gravitation and its relations to the other forces of nature.

The general theory explains gravity beyond Newtonian physics. Newton explained that the mass of a body causes gravity and showed how to calculate the force. But Einstein explained *how* matter causes gravity. To understand the general theory, we have to go back to the frames of reference. Einstein asked what nature has to do with these frames of reference and their respective states of motion and concluded that laws of motion must be entirely independent from the choice of the particular reference frame used to describe nature. This point—*that the laws should be entirely independent of the choice*—is called the general principle of relativity.

His search for unity and simplicity in scientific principles led Einstein to this theory, the same objective that first led him to the special theory and the reason he later pursued the unified field theory. As illustrated earlier in this chapter, application of the special theory is limited to frames of reference moving in a straight line toward or away from each other at a constant speed—that is, without acceleration. The general theory, on the other hand, provides a formula for the relationship of matter *throughout space moving in any direction, with or without acceleration*. It provides a unifying and simplifying principle that (1) modifies the principle of the constant velocity of light to address encounters with strong gravitational fields and (2) accounts for the fact that all bodies (regardless of mass or composition) fall freely in a gravitational field with the same acceleration.

The main predictions derived from the general theory include these:

1. All electromagnetic radiation (including light) is deflected by gravitational force.

2. Mercury's orbit deviates from the orbit calculated by Newtonian physics.

3. A clock on the surface of a massive object will run slower than an identical clock in free space.

4. Gravitational *waves* exist, radiating at the speed of light, from large masses that are accelerating.

Numbers 1, 2, and 3 have been proven, while number 4 is the subject of intensive investigation by a large community of scientists at universities and research centers throughout the world. The end result of general relativity is a further extension of physics to account for gravity and accelerated frames of reference and leads to the conclusion that space is "curved" and that not only are space and time linked but the complete explanation of the universe combines space, time, *and matter*. In 2011, scientists at NASA published a report titled "At Long Last, Gravity Probe B Satellite Proves Einstein Right." The satellite orbited earth for fifteen months, using gyroscopes to measure a central aspect of the general theory of relativity—namely, the assertion that *gravity appears when mass bends space and time.* The scientists reported that "the rotating Earth twists spacetime much as turning a heavy bowl twists a table cloth beneath it."

Relativity Impacts Our Everyday Lives

An Aid to Navigation

As obtuse and esoteric as relativity is, it clearly has practical impacts on our lives. Einstein's equation $E = mc^2$, which grew out of the special theory, opened the way for atomic research. Further, the construction of particle accelerators requires an understanding of special relativity's mass/energy principles. In turn, the field of nuclear medicine greatly depends on the creation of isotopes in particle accelerators. Nuclear medicine is the branch of medicine that uses radioactive materials for the diagnosis and treatment of disease. Special relativity has a direct impact in this area. For instance, the element technetium is widely used in bone scans for cancer patients to determine whether the cancer has metastasized to the bone. Technetium does not exist in nature, but was first discovered in 1937 as a result of high-energy radiation of the element molybdenum in a nuclear particle accelerator.

General relativity's ties to modern technology can be seen in the fast-growing use of the global positioning system (GPS), which is a precise satellite-based navigation and location system originally developed for U.S. military use and now is used in multiple conflicts around the world. But GPS has a wide variety of existing and potential nonmilitary uses, including the system many of us use every day in our personal autos, personnel tracking devices, and smartphones. GPS is used to detect the slow movements of tectonic plates over generations and to identify the exact location of sea and land travelers. To maintain such precision, GPS must incorporate adjustments to the satellites' locations and measurements that are calculated based on Einstein's general theory of relativity.

In 1919, four years after Einstein had completed his work on the special and general theories of relativity, he wrote:

Let no one suppose . . . that the mighty work of Newton can really be superseded by this or any other theory. His great and lucid ideas will retain their unique significance for all time as the foundation of our whole modern conceptual structure in the sphere of natural philosophy.

Einstein modestly described the special theory of relativity as "simply a systematic development of the electrodynamics of Maxwell and Lorentz," which put the "finishing touch to [their] mighty intellectual edifice." He also recognized the importance of the discoveries of Marie Curie and Ernest Rutherford, through which he became aware of the energy in matter. This, in turn, led him to $E = mc^2$. "We see atom disintegration," said Einstein, "where nature herself presents it, as in the case of radium."

Beginning with the moment he posed the innocent question at the age of sixteen, in 1895—*What would the world look like if I were riding on a beam of light?*—and culminating with the full development of the general theory of relativity in 1916, Einstein brought together the diverse concepts and discoveries of his predecessors, like discordant notes of musical instruments, until he formed a full symphony, a revolution in physics, that exploded with a mighty crescendo heard around the world. Thus, like Newton, Einstein stood on the shoulders of giants, and like Newton, his greatness exceeded those who came before him.

Though Einstein's fame rests on relativity, he wrote prolifically on numerous other scientific subjects in his early years. In the ten years between his two major publications on relativity, he wrote sixty papers on such topics as photochemistry, quantum statistics of gas, thermodynamics, optics, molecular physics, electromagnetism, and the history and philosophy of science. Between 1915 and 1930 his work on quantum theory was highly influential. All together, his scientific publications total about 350.

On April 18, 1955, after a life that included two of the greatest contributions to twentieth-century physics and that was intertwined with the rise and fall of Germany, the formation of Israel, and creation of the atomic bomb, Albert Einstein died in his sleep at Princeton Hospital. Of the two passions that dominated his existence, he once said, "Politics are for the moment. An equation is for eternity."

PART FOUR

The Big Bang and the Formation of the Universe

In Focus

From Galileo's observations beginning in 1610 we gained but an inkling of the vastness of the heavens. The full story didn't unfold until the twentieth century when giant telescopes and other equipment allowed us to see the true size and nature of the galaxy that is Earth's home, then to see galaxies beyond it.

Our solar system is located in the outer reaches of a vast rotating galaxy—the Milky Way—and moves in relation to its center at a speed of 140 miles per second. One revolution of this group of 300 billion stars takes 230 million years. The Milky Way Galaxy is but one of billions of galaxies, all of which are still reeling from a giant explosion called the Big Bang, which continues to show its effects today in the light spectra of distant galaxies and in the actual cosmic radiation that remains measurable throughout the universe. By measuring the speed of the galaxies' recession we've been able to calculate that the entire universe formed about 14 billion years ago—not only the formation and motion of the galaxies but also the creation of space-time itself.

Through a detailed understanding of the elementary particles that make up atoms and by studying the cosmos, we've been able to determine the conditions that existed immediately after the explosion and the subsequent events that evolved from it. One result was the formation of the elements that make up Earth and its inhabitants. Thus we can look back and see the development of both the physical universe and life on this planet.

Among many questions that remain to be answered are two of immense proportions: Where did the "material" that exploded in the Big Bang come from? What is the ultimate fate of the universe? Part Four examines the discovery of the Big Bang and its implications, and explores possible answers to those two monumental questions.

Chapter Nine
THE COSMIC EGG

I have observed the nature and material of the Milky Way. With the aid of the telescope this has been scrutinized so directly and with such ocular certainty that all the disputes which have vexed philosophers through so many ages have been resolved. . . . The galaxy is, in fact, nothing but a congeries of innumerable stars grouped together in clusters. . . . And what is even more remarkable, the stars which have been called "nebulous" by every astronomer up to this time turn out to be groups of very small stars. . . . In the nebula called the Head of Orion, I have counted twenty-one stars. The nebula called Precipe . . . is a mass of more than forty starlets.

—Galileo Galilei, *The Starry Messenger* (1610)

On the far side of the Big Bang is a mystery so profound that physicists lack the words even to think about it. . . . The Universe consisted of . . . maybe only one particle that interacted with itself in that tiny, terrifying space. Detonating outward . . . every bit of matter was crushed with brutal force into every other bit, within a space smaller than an atomic nucleus.

—Robert Crease and Charles Mann, *The Second Creation* (1986)

Look up to the heavens. At night, about 5,000 stars and a few planets are visible to the naked eye. Until about four centuries ago, this seemed to be the extent of the universe—a finite number of unchanging immovable stars spread evenly throughout a finite area. Except for the occasional nova like the one Tycho Brahe observed in 1572, the universe of stars surrounding our solar system appeared much as Aristotle had described it more than 2,000 years ago. But, as we saw in Chapter 2, in 1609 Galileo improved the telescope and began to study the band of constellations that included Orion,

Perseus, Cassiopeia, Cygnus, Aquila, Sagittarius, Centaurus, and Carina. This band was likened to a river of milk by the ancient Greeks because of its light, cloudy appearance, and was later called the Milky Way. Galileo saw that the Milky Way was not composed of "clouds" at all. It was a vast collection of stars. Thus began the Renaissance revolution beyond Copernican theory and the solar system, as it extended to the starry heavens. Centuries later, this revolution not only revealed what exists in the heavens but also unveiled an extraordinary and intricate pattern and story that no one could have ever imagined—a story that tells how the heavens came to be the sun, Earth, planets, and stars, and the very matter from which life formed.

Astronomers Discover Humankind's Galaxy and Location in It

In and Out of Spiral Arms

In 1781, the self-taught British astronomer Sir William Frederick Herschel (1738–1822) discovered the planet Uranus, the first planet to be discovered for centuries. Then in 1785, after carefully cataloging the positions of thousands of stars, Herschel was first to suggest that our solar system is actually part of a larger system of at least several million stars and that it has the shape of a "thin disk of nearly infinite extension." After twenty more years of systematic observation with the most powerful telescopes in existence, which he constructed himself, Herschel cataloged 2,500 star clusters in the Milky Way Galaxy.

The quality of telescopes continued to improve dramatically in the 1800s, enabling other astronomers to see millions of stars never before observed, allowing for improvements in measuring the distance of stars from Earth, and leading to the discovery of the eighth planet, Neptune, in 1845. In the early decades of the twentieth century, based on work performed by Harlow Shapley (1885–1972) and Robert J. Trumpler (1886–1956), we gained our modern understanding of the Milky Way Galaxy. Their work demonstrated that our solar system is located in a spiral arm near the edge of this giant rotating system of stars, which measures about 100,000 light-years across. The center of the galaxy is located approximately 27,000 light-years away

FIGURE 9-1
Milky Way Galaxy

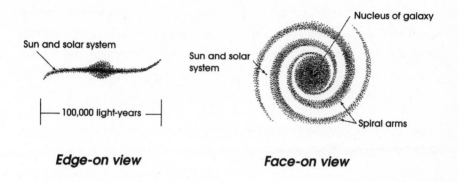

Edge-on view   **Face-on view**

from us. *That is, traveling at 186,000 miles per second (the speed of light, tra-versing about 5.88 trillion miles in a year), it would take 27,000 years to get from Earth to the center of the Milky Way Galaxy.*

As astronomers continued to map the heavens, they began to realize the incredible vastness in which we live and that William Herschel's estimate of at least several million stars in the Milky Way was a gross underestimate. We came to see that in our Milky Way Galaxy alone there are actually about 300 billion stars. Of them, we are relatively "near" to only one. Our sun is 93 million miles from Earth. It is an average-size star in the center of our solar system but accounts for 99 percent of the total mass of the solar system. Our solar system of planets and other orbiting material (comets, asteroids, dust, gas, and other space debris) is located in one of the spiral arms and orbits the center of this gigantic mass of stars, together with the 300 billion other stars and solar systems of the Milky Way. The star closest to our solar system is called Proxima Centauri, and is located about four light-years away (23 tril-lion miles), in the same spiral arm. The galaxy also contains enormous hy-drogen and dust clouds that astronomers are still in the process of detecting and measuring. This material extends so far beyond the visible stars that if it were included as part of the Milky Way, the galaxy would be about 500,000 light-years across, rather than the 100,000 light-years measured without the clouds.

Our solar system and this area of the spiral arm move at a speed of 140 miles a second—that is, about 500,000 miles per hour in relation to the center of the galaxy. One revolution takes about 230 million years. In the lifetime of the universe, our sun has made twenty revolutions around the center of this immense galaxy. If we were to view this rotation from a distance and speed it up, we'd see the galaxy pass through various stages in this ponderous motion and evolution. Just as the planets revolve about the sun with the more distant planets moving slower than Earth, Venus, Mars, and Mercury, the outer areas of the Milky Way rotate progressively more slowly.

Astronomers Discover a Hundred Billion More Galaxies

The Starry Messenger Revisited

Is this vast revolving cluster of 300 billion stars called the Milky Way the entire universe? Does this collection of burning hydrogen and helium spheres, along with Earth and some planets and other matter void of life, comprise all that exists? Is there only infinite space beyond?

In 964 CE, the Arab astronomer al-Sūfī wrote *Book of the Fixed Stars*, and was the first to note a small object that looked like a fuzzy star. In the year 1612, the German astronomer Simon Marius (1573–1624), using the newly invented telescope, rediscovered this faint area far out in space, saying it looked like "the light of a candle seen through a horn." It came to be called the Andromeda Nebula, and was believed to be a luminous cloud of gases and dust in the Milky Way Galaxy. Despite subsequent improvements to the telescope, individual stars in the Andromeda Nebula were not visible until 1885, when a nova appeared in Andromeda for several days. (A nova is a massive nuclear explosion in a star.) The nova indicated that Andromeda consisted of more than just dust and gases—at first, it was believed to be a *cluster of stars* on the distant fringes of the Milky Way.

American astronomer Edwin Powell Hubble (1889–1953) was the founder of extragalactic astronomy, and his namesake, the spectacular Hubble Space Telescope launched in 1989, pioneered research on Andromeda. After earning degrees in math and astronomy in 1910 at the University of Chicago, he studied law as a Rhodes Scholar at Oxford and began practicing law in Kentucky in 1913. Becoming bored with the legal profession, Hubble abruptly changed directions once again and earned a PhD in astronomy at the University of Chicago in 1917. He later went to work at Mount Wilson Observatory in California, where the new 100-inch telescope finally made it possible to see the vast number of individual stars in Andromeda.

Hubble's work at Mount Wilson showed that Andromeda was not just gases and dust and a few novas; it was made up of stars, billions of them. Even more amazing, it wasn't in the Milky Way at all; it was far, far outside our galaxy. By 1923, Hubble had determined that Andromeda is another galaxy in its own right, an independent spiral galaxy about 200,000 light-years in diameter and about *2 million light-years* from the Milky Way. Andromeda was the first galaxy proved to be beyond the Milky Way. The *nearest* large galaxy, Andromeda is the *farthest* object that can be seen with the naked eye. Figure 9-2 is a photograph of Andromeda and its two companion galaxies (called NGC 205 and M32) taken with the forty-eight-inch telescope at Mount Palomar Observatory northeast of San Diego, California.

Hubble further determined that thousands of other "nebulae" are also

FIGURE 9-2

The Andromeda Galaxy

NASA/JPL/California Institute of Technology

galaxies, much like Galileo had written of the Milky Way nebulae more than 300 years earlier. Thus the map of the universe, and humankind's concept of where we fit in that universe, had to be redrawn once more. *Since 1925, astronomers have found that there are about 100 billion separate galaxies in the observable universe, averaging 100 billion stars each. The farthest* galaxy from Earth, known as 4C41.17, is about 15 billion light-years away. The *largest* discovered so far (called Abell 2029) is about sixty times the size of the Milky

FIGURE 9-3
Galaxy Shapes

European Space Agency & NASA

Spiral galaxy Messier 101, located in Ursa Major constellation (aka the Big Dipper).

| Elliptical | Spiral | Barred Spiral |

Hubble classification scheme of galaxies.

Way, with over 100 trillion stars and a diameter of 6 million light-years. Like the Milky Way, most of these groups of stars are shaped like whirlpools of light, disc-like structures with spiral arms. About 90 percent of all the galaxies in the universe can be classified into one of three broad categories (shown in Figure 9-3): elliptical, spiral, or barred spiral.

There are single galaxies, galaxy pairs, and large galaxies with smaller satellite galaxies. Most galaxies are part of "clusters" of galaxies, averaging about 100 galaxies per cluster. In the constellation Coma Berenices there is a cluster of about 10,000 individual galaxies. The Milky Way is part of a cluster of 20 galaxies. The clusters are distributed throughout space in groups that form a generally uniform or homogeneous pattern, with an average of about 10 million light-years between each group.

Hubble Calculates the Recession Speed of the Galaxies as the Universe Rushes Apart

Sirius and the Red Shift

In the late 1800s, astronomers and physicists began to develop a method to determine the movement of stars and other heavenly bodies toward or away from Earth by measuring the light spectra of such bodies as we perceive their light here on Earth. London astronomer Sir William Huggins (1824–1910) and French physicist Hippolyte Fizeau (1819–1896) discovered that certain spectral lines of some of the brighter stars are "shifted" slightly to the red from their normal position in the light spectrum. They correctly concluded that this was similar to the Doppler effect (named after the Austrian physicist Christian Doppler) that pertains to sound waves. The Doppler effect explains the phenomenon of sound waves being "compressed" together when the source of the sound (such as a train whistle or ambulance siren) is moving toward the listener—causing an increasingly higher pitch—and the waves being "stretched" apart when the source is moving away from the listener, causing an increasingly lower pitch. Although light is made of photons and sound results from air vibrating, the two are similar in the sense that both come in wavelengths that can be measured. *A change in the light wave toward the red (lower frequency or longer wavelength) results from the star moving away from the observer on Earth.* In 1869, Huggins determined that Sirius (the brightest star in the northern sky) is moving away from the Earth at about twenty miles a second.

Edwin Hubble took this information about the shift in light spectra and made his second remarkable discovery, one that would turn out to be the greatest discovery in astronomy of the twentieth century. In 1927, combining the previous research on light spectra with the powerful new telescopes at Mount Wilson, Hubble discovered that the red shift of the receding galaxies increases in proportion to the distance from us. In other words, the universe is expanding with the farthest stars moving faster, and the rate of expansion can be represented by his calculation that is now called Hubble's constant: *recession speed equals distance divided by a time factor of about 15 billion years. Every galaxy cluster (but not necessarily every galaxy) is rushing away from every other cluster of galaxies, precipitously, at varying velocities in accordance with Hubble's constant, with the galaxies at greater distance receding at a faster rate.* Today's calculation of Hubble's constant shows that the galaxies are expanding at a velocity of 10 to 20 miles per second for every million light-years in distance from Earth. For example, the Ursa Major II cluster of galaxies is receding at 25,930 miles a second, 14 percent of the speed of light. In the universe's outermost regions, galaxies have been clocked at 84 percent of light speed.

The expansion discovered by Hubble is of the space *between* the clusters of galaxies, not *within* galaxy clusters. Each galaxy and galaxy cluster is bound together by gravity and is not expanding. So, for example, the Andromeda Galaxy and the Milky Way, part of one cluster in which galaxies are gravitationally attracted, are actually moving *toward* each other and will collide or coalesce hundreds of millions of years from now. Throughout the universe in general, however, a rapid expansion is taking place. As we discuss in the next several pages, there is an inescapable connection between that expansion and the very beginning of the universe.

By Calculating the Universe's Expansion in Reverse, We Are Led to the Cosmic Egg and the Big Bang

Zero-Time at a 100 Trillion Degrees

Until the last century, theories about the beginning of time tied the origin of humans and all living things to the formation of the universe. Irish bishop James Ussher (1581–1656) wrote a chronology of the Old Testament, in which he added up all the generations of men and women mentioned in the Bible since Adam and Eve, and pegged Creation at 4004 BCE, *at 2:30 p.m., Sunday, October 23!* One person is said to have asked the bishop, "And pray, Holy Father, what was God doing *before* he created the universe?" To which Ussher replied impatiently, "Creating hell for those who ask such questions!"

An understanding of the receding galaxies through twenty-first-century astrophysics has allowed us to look back in time to the actual formation of the universe in an event that Bishop Ussher could never have dreamed possible. If we imagine the expansion of the universe in reverse and mathematically calculate its motion in reverse, all the galaxies would simultaneously meet at a point. This is called "zero-time" and considered the beginning of the universe. Although there is still some question about how long ago that was, most experts agree that zero-time was about 14 billion years ago.

"Once upon a time there was no time," wrote John Barrow in *The Origin of the Universe.* What happened at zero-time and thereafter to bring us to where we are today? In 1927, after becoming aware of Hubble's discovery of the expanding universe, the Belgian astronomer Georges Édouard Lemaître (1894–1966) proposed the theory that is now generally accepted by astronomers and other experts. At zero-time, he said, all matter was in one tiny mass that he called the "super-atom" or the "cosmic egg." Nothing else existed—no galaxies, planets, stars, or life. Indeed, the elements didn't exist because the cosmic egg was composed of energy not yet even in the form of nuclei or atoms. Without matter, there was no time. The cosmic egg was subject to its own gravitational pull and contracted and grew closer and closer together, creating higher and higher temperatures as it was compressed into a smaller and smaller volume. At that moment, the entire universe was a "kernel" of energy. At some point, with intensely high temperature at the smallest volume possible, a great explosion occurred, and this single kernel of energy became everything there is. Lemaitre said that the recession of the galaxies is the visible evidence of this explosion. His theory also explained the recession of the galaxies within the framework of Einstein's general theory of relativity.

The explosion is the earliest event about which we have any record. The theory was further refined and modified by George Gamow (the Russian-born physicist responsible for the "liquid droplet" model of the atomic nucleus, discussed in Chapter 6) and published in a 1948 paper titled "The Origin of Chemical Elements," in which he postulated that the elements were formed by atomic nuclei being built up by the successive capture of neutrons and in which he first coined the term "Big Bang." The Big Bang theory was then made popular in 1950 by the British astrophysicist Fred Hoyle in a series of BBC radio talks titled *The Nature of the Universe*. It is so widely accepted that astronomers now call it the "standard model."

We now have a surprising amount of reliable circumstantial evidence about the Big Bang. Steven Weinberg (1933–), a physics professor at the University of Texas, Austin, received the 1979 Nobel Prize in physics for his work in this field. He described the results of that research in his book, *The First Three Minutes*:

> In the beginning there was an explosion. . . . At about one hundredth of a second . . . the temperature of the universe was about one hundred trillion degrees centigrade. This is much hotter than in the center of even the hottest star, so hot, in fact, that none of the components of ordinary matter, molecules, or atoms, or even the nuclei of atoms, could have held together. Instead, the matter rushing apart in this explosion consisted of various types of the . . . elementary particles. . . .
>
> As the explosion continued the temperature dropped . . . reaching one trillion degrees at the end of the first three minutes. It was then cool enough for the protons and neutrons to begin to form into complex nuclei, starting with the nucleus of heavy hydrogen . . . which consists of one proton and one neutron. The density was still high enough . . . so that these light nuclei were able to rapidly assemble themselves into the most stable light nucleus, that of helium, consisting of two protons and two neutrons. . . . At the end of the first three minutes the

contents of the universe were mostly in the form of light, neutrinos and anti-neutrinos. . . . This matter continued to rush apart, becoming steadily cooler and less dense. Much later, after a few hundred thousand years, it would become cool enough for electrons to join with nuclei to form atoms of hydrogen and helium.

Although the Big Bang occurred about 14 billion years ago, it took several billion more years for the galaxies to acquire their present configuration in the universe, including the Milky Way and our own solar system within it. The hydrogen and helium atoms rotated increasingly faster, with most of that mass flattening along the axis of rotation. Like the Milky Way, most of this hydrogen and helium formed into disc-like structures with spiral arms—great wheels of matter and light—as well as the other galaxy configurations shown in Figure 9-3.

Here, in the outer reaches of the Milky Way Galaxy, about 4.5 billion years ago a cloud of hydrogen and helium began to condense and fall inward upon itself because of its own gravitational pull. The matter from which it was formed could have come from the subsequent explosion of a supernova, rather than forming at the Big Bang. In either case, this matter first formed a flat, spinning disc, like the Milky Way itself. But as more and more material collected, the center of the disc grew thicker, and the greater gravitational pull from the enlarging object caused more space debris to be attracted until the object formed into the most compact geometric shape—the sphere—and became so massive that its interior was under incredible physical pressure. That pressure caused the new sphere to heat to temperatures up to 20 million degrees Fahrenheit, which caused the hydrogen atoms to fuse into helium in a continuous and massive series of nuclear reactions, marking the birth of our sun. This process was repeated in the birth of every one of the hundreds of trillions of other stars in the universe. As discussed in Parts Two and Three, when this fusion takes place we see—and we exist as the result of—the transformation of mass into light and heat.

The original explosion might have catapulted this material with sufficient energy to escape from the gravitational pull of the rest of the material. If it did, the universe will expand forever, as most scientists believe and is described below. On the other hand, if the material is not traveling with enough energy, all of the universe will fall back on itself someday and perhaps create another Big Bang and begin the process all over again. Even if this were to occur, it would be tens of billions of years before the universe meets that ultimate fate.

Heavier Elements Coalesce to Form 1 Percent of the Universe, Including Earth

Pluto and the Oort Cloud

As Steven Weinberg explained, we now understand the process that led to the formation of the elements during and after the Big Bang. This event, so difficult for the average human mind to envision, produced space, time, light, forces of nature, and fundamental particles. The cosmic egg predominantly formed into hydrogen atoms (which contain only one proton and one neutron), followed by the next simplest atom, helium (with only two protons and two neutrons). These two elements make up a bit less than 99 percent of the entire universe (74 percent hydrogen and 25 percent helium). The Big Bang produced dark energy (about 74 percent of the mass of the entire universe) as well as dark matter (about 22 percent), and over the succeeding hundreds of millions of years, the heavier elements that compose the remaining approximately 4 percent of baryonic, or ordinary, matter in the universe formed as an offshoot of early stars' fusion reactions and as a result of catastrophic stellar explosions. In a relative sense, there has been a very, very small nuclear change or change in the elemental makeup of the universe from the time immediately after the Big Bang until now. Trillions times trillions times trillions of interactions of hydrogen atoms, helium atoms, and other elementary particles occurred to form elements other than hydrogen or helium—*yet, those ninety-eight other naturally occurring chemical elements make up only about 4 percent of the entire universe.*

Focusing for a moment on *living matter on Earth*, it is also remarkable that 99 percent of the material making up all living organisms is made of just *four* atoms: hydrogen, carbon, nitrogen, and oxygen. It's likely that every atom in your body was once part of a star. To form carbon, for example, three helium nuclei typically collide with each other simultaneously in a millionth of a millionth of a second. With a couple of relatively minor exceptions to this process, the carbon atoms in every organic molecule in every cell in every living creature on Earth was formed by such a triple collision— part of the spectacular transformation from elementary particles to living beings that can now comprehend from where we came.

As mentioned earlier, matter gravitationally "pulls" itself into spheres because that geometric figure possesses the smallest surface (that is, the most efficient configuration) for a given total volume of matter. Thus, around the same time as the sun was forming, dust and gas that included heavier elements were floating through our galaxy and began to coalesce into larger masses attracted to the sun by its enormous gravitational pull. As the sun condensed, so did these other masses of matter, eventually forming more solid bodies, which collided with each other over a period of millions of years and combined into larger solid spheres. This coalescing of matter and succession of collisions formed planet Earth, the other planets, and the numerous moons of planets in our solar system.

Initially, the Earth was extremely hot and had no atmosphere. An early atmosphere was then formed by emission of hydrogen sulfide and other gases from the molten materials. Ten billion years after the Big Bang the Earth's primordial soup gave rise to the first organic molecules. In 1995, after a decade of planning and traveling to its destination, a 746-pound satellite probe from the $1.35 billion *Galileo* spacecraft entered Jupiter's atmosphere and radioed a wealth of data back for 57 minutes before it plunged into the planet and was crushed and melted by Jupiter's immense gravitational force and heat. This measurement of oxygen, hydrogen, and other gases and temperatures by the probe's ultraviolet spectrometer gave us important information about *Earth's* early moments of formation billions of years ago because Jupiter is so massive and its gravity so great that its atmosphere is virtually the same as when it and the entire solar system was formed 4.5 billion years ago. Those same gases formed the basis for Earth's origin, but were gradually altered by their escape to space, bombardment of meteors and comets, chemical reactions, and volcanic activity. In contrast, although Jupiter has undergone its own pressure and chemical changes, it remains a relatively pristine example of the original primordial cloud and provides valuable clues to Earth's evolution through comparison of the two planets.

Enormous amounts of debris in our solar system never coalesced into planets or moons—these are the asteroids, comets, and other dust and material that are particularly common around the outer planets. Asteroids are made of rock and metal and come in all sizes, from a few ounces to millions of tons, and the vast majority (about 7,000) orbit the sun in an asteroid belt that exists between Mars and Jupiter. The largest known asteroid (Ceres) is 581 miles in diameter. Pieces of asteroids result from collisions between them, and a number of those pieces strike the Earth every year in the form of meteorites. Comets, on the other hand, are primarily composed of water, ice, hydrogen, carbon, oxygen, and nitrogen and number in the trillions in our solar system. This material forms a "tail" of vapor, gas, and dust; hence the term *comet*, from a Greek word meaning "long-haired." Nearly all comets revolve around the sun in a plane similar to the plane of the major planets, and often cross planets' paths, as seen with Halley's comet. Some comets' orbits are short (within the area between Earth and sun), while others have tremendously elongated paths that extend far beyond Pluto's orbit. Most astronomers subscribe to the theory that comets originate in a vast cloud of material called the Oort Cloud, named after the Dutch astronomer Jan Hendrick Oort (1900–1992) who first proposed the theory of the existence of this material in that region and how it results in comets. In 2004, the Stardust Mission flew by a comet named Wild 2 and captured particles from the tail of dust and gas (the coma) that forms when a comet approaches the sun. Using mineralogical and isotope data from these particles, scientists reported in 2012 that one of the fragments, which they called "Iris," originated 3 million years after the formation of the earliest solids in our solar system. Because transport of solid materials across the solar system is most likely to have occurred before Jupiter formed, these results support the concept of massive Jupiter also having formed 3 million years after the earliest solids.

Astronomers Discover Other Solar Systems in the Milky Way Galaxy

Search for Life on Mars

Beginning in the early 1980s, improved methods to search for other solar systems have led to the discovery that planets orbit other stars in the Milky

Way. The first such discovery took place in 1992, when astronomers found a star with two planets 1,300 light years from Earth. In October 1995, astronomers detected a planet about half the size of Jupiter orbiting a star called 51 Pegasi in the constellation Pegasus. In January 1996, two California astronomers detected two planets larger than Jupiter accompanying the stars 70 Virginis, in the constellation Virgo, and 47 Ursae Majoris in Ursa Major (the Big Dipper), about thirty-five light-years away. Trillions of miles from Earth, these planets are too distant to be seen with telescopes, but sensitive instruments (such as the Infrared Astronomical Satellite and powerful Earth-based devices) can monitor stars to see if they wobble, indicating gravitational pull by a planet. These instruments can also measure the pattern of infrared light around the thin, gaseous, pancake-like disc around the star to detect the ring-shaped gap that any revolving planet would sweep out.

The temperatures of the planets in Virgo and Ursa Major are between 100°F below zero and 185°F above zero—a range conducive to the chemical processes that could allow for life to be formed and sustained. Researchers have also found ordinary vinegar in a stellar cloud 25,000 light-years from Earth, by analyzing the unique spectrographic "signature" of its wavelengths of light and radio signals from space. Vinegar consists of an organic molecule (acetic acid) that might have played a role in the formation of life on Earth. The same technology has been used to identify all four elements critical to the formation of life on Earth—hydrogen, carbon, nitrogen, and oxygen—in massive clouds of dust and gas throughout the universe.

The speculation about life elsewhere in the universe took a startling twist in August 1996 when NASA announced the discovery of fossilized organic molecules in a meteorite from Mars. The rock had been catapulted from Mars about 15 million years ago when a huge asteroid collided with the planet. It landed on Earth about 13,000 years ago, and contains polycyclic aromatic hydrocarbons (PAHs), potential evidence for the *presence of biological organisms* that produced the PAHs about 3 billion years ago—that is, about the same time that replicating (living) organisms were first forming on Earth. In 2012, the presence of PAHs as "carbon from Mars" was confirmed by Raman and mass spectroscopy, as reported in the May issue of *Science* magazine. In 2010 and 2011, the presence of water on Mars was largely confirmed. Astrobiologists are fascinated by clay minerals because generation of clays requires long-term interactions of rock and water under environmental conditions that typically support organic-based "life." Spectrometers aboard NASA's *Mars Reconnaissance Orbiter* and the European Space Agency's Mars Express mission have proven the existence of numerous water-generated deposits of clay on Mars. And ice has been found buried in Martian soil along with thick glaciers. In 2012, NASA's *Curiosity* rover landed on Mars and reached a clay-rich target, the huge Gale crater on Mars. The images taken by the rover's cameras show strong evidence for torrents of water and salty puddles. As a result of such discoveries, the search for primitive life forms produces tremendous enthusiasm among scientists who are thinking that recognizable microorganisms will eventually be found. These simple organisms, whether they will have separate origins, a distinctive biochemistry, or are related to life on Earth (see Chapter 18), could open important new windows to understanding the origins of us all.

In light of the existence of trillions of stars, the recent discoveries of planets, and the possibility of past life on Mars, it appears that our solar system and perhaps even planet Earth are not unique. It seems highly unlikely that one ordinary star, our sun, is the only star accompanied by a planet that sustains life. In fact, astronomers have now used the Kepler Space Telescope to find a planet in the midst of what is known as "the habitable zone," in which temperatures are found to be suitable for life as we know it. Orbiting a sun-like star, planet Kepler-22b is 600 light-years from Earth and is about 2.5 times its diameter. The scientists expect the surface (if it has a surface) to be about 70°F. The Kepler telescope has been used to identify more than 2,000 new planets in the Milky Way Galaxy.

If life or past life is discovered on Mars or elsewhere, humanity's sense of itself will be shaken once again. Amid the developing technology, astronomers are in the dawn of a new era, the second coming of Columbus wherein new worlds are discovered every day. Perhaps there are millions of life-sustaining bodies out there. In those planetary systems that we cannot see, but that exist, intelligent life might have evolved. Not only are we dealing with immense numbers of stars and potential solar systems but we must also consider this in the context of the 14 billion years during which the universe has existed. Living systems, millions of them, could have evolved on other planets and could be scattered throughout the universe, some overlapping in their evolution and stages of development, with the birth of others coming millions of years after the death of ones preceding. After all, there was a time when Earth itself was totally inhospitable to any form of life. The environment evolved and the planet gave birth. Our time as a product of this changing planet is less than a second in infinity. As we will see in Part Six, under proper conditions, life could form anywhere in the universe. Life on another planet could have its time, too—entirely different forms of life suited to that planet alone, passing through its billion-year cycle, then becoming inhospitable, desolate, dead once more. A billion planets could have their cycles—and no living, thinking being from one planet would know first-hand about another.

As suggested by the question asked of Bishop Ussher, without cause there is no effect. Despite Aristotle's belief that the universe always existed, there is no physical phenomenon in our knowledge and experience of the universe that proves that energy or matter can come about spontaneously. Based on Einstein's equation, $E = mc^2$, we are reminded that mass and energy are interchangeable. Neither energy nor matter can be created from *nothing*. Because it is clear that the universe is expanding at an increasing rate, with all matter moving away from all other matter, the universe could not have existed forever in its current state. That is, the universe can't be infinitely old because it would have already expanded beyond existence (see further discussion following). Therefore, the universe must have come into existence a finite time ago. Thus the question must be: Where did the cosmic egg come from? What was the "First Cause"? How did Lemaître's super atom—that primordial egg of incredible density and ultimate temperature—come to be in the first place, and why did it explode?

The implications of these questions are immense. Perhaps someday a scientific theory will provide the answers. Some scientists have suggested that the cosmic egg was in existence for eternity before the Big Bang. Other people postulate that the cause of the cosmic egg or the cosmic egg itself was a supreme being or other supernatural force. Some cosmologists have developed theories that don't rely on the common understanding of cause and effect because the same physical laws don't apply when there is no time, no space, and no matter.

But assuming the usual meaning of *cause and effect*, we could easily conclude that such a supernatural force *must* have created the "stuff" that made up the cosmic egg. There is ample room in astrophysics for a supreme being at creation. But then, is there any other compelling or inescapable conclusion or belief that results from this fact or assumption? That is, even if everyone agreed that a supernatural force was the cause of the cosmic egg, are any further empirical conclusions derived from that "fact"?

The various creation myths of ancient times attempted to provide an explanation for our existence, to remove the unknown and provide psychological security and stability to society. Indeed, sociologists and psychologists have determined that primitive and medieval societies that were homogenous in thoughts, goals, and particularly religion, were most likely to succeed and thus survive in a true Darwinian sense. The societies most likely to prosper were the ones to pass on their mores and beliefs more effectively, just as the fittest individuals of a species are the most successful at passing on their genes to their offspring (see the next section). Religion and other belief systems provided a survival advantage to societies when very little of the natural world was understood. Even though the Big Bang and the time it occurred have been proven well beyond debate, millions of people continue to subscribe to the biblical precepts of spontaneous creation of the universe by God about 5,800 years ago. The evidence supporting the Big Bang is overwhelming. In light of that evidence, belief in the eternal "truths" about creation that were conjured up centuries ago is far more dangerous to the progress and well-being of humankind and far less rational than it was when people allowed Aristotle's and Ptolemy's stories to be carried with such force into the Renaissance.

Beginning immediately after the first instant of creation, the universe evolved in a predictable fashion. Thus even if one assumes that a supernatural being or force was responsible for creation of the universe at the time of the Big Bang, divine intervention is no longer needed for the long list of questions that we began to ask and have now answered in our more recent history. In *A Brief History of Time*, Stephen Hawking puts it this way:

> The whole history of science has been the gradual realization that events do not happen in an arbitrary manner, but they reflect a certain underlying order, which may or may not be divinely inspired. . . . These laws may have originally been decreed by God, but it appears that he

has since left the universe to evolve according to them and does not now intervene in it.

Unresolved mysteries persist. But they are contained within a resolved framework. From this small planet in the outer reaches of an unremarkable galaxy, men and women have examined the vast space that surrounds them and unraveled a most magnificent description of the origins of the universe and life on Earth.

Chapter Ten
THE ECHO OF THE BIG BANG

The Earth is a place. It is by no means the only place. It is not even a typical place. No planet or star or galaxy can be typical, because the Cosmos is mostly empty. The only typical place is within the vast, cold, universal vacuum, the everlasting night of intergalactic space, a place so strange and desolate that, by comparison, planets and stars and galaxies seem achingly rare and lovely.

— Carl Sagan, *Cosmos* (1980)

The first chapters of the Big Bang theory were written by astronomers whose great telescopes explored the sky during the first few decades of the last century. Then a startling and unexpected new chapter was written by those in a new field of astronomy—the radio astronomers—when they discovered complex clues scattered throughout the universe from 14 billion years ago. Not only can we *see* the evidence of the Big Bang through the recession of galaxies, we can actually measure the remnants of the "heat" left over from that explosion.

As we learned in the previous chapter, the early universe was incredibly hot. The exceedingly high temperature was significant not only because of its effect on atomic particles but also because the entire universe was bathed in electromagnetic radiation, which *cooled* along with the matter. To understand how the detection of this radiation led to the confirmation of Hubble's observations and Lemaître's Big Bang theory, we must first define electromagnetic radiation.

Maxwell Predicts the Existence of Invisible Waves throughout the Universe

Volts, Amps, and Coulombs

As suggested by the term itself, *electromagnetism* refers to one force or phenomenon that is a combination of electricity and magnetism. Following Newton's example of applying the scientific method to the branch of physics called mechanics, eighteenth-century researchers like the French physicists Charles Coulomb (1736–1806) and André-Marie Amperè (1775–1836), the Italian physicist Alessandro Volta (1745–1827), and the German mathematician Carl F. Gauss (1777–1855), performed hundreds of experiments with electricity and magnetism in an effort to understand these phenomena. Their work led to an understanding about the nature and characteristics of electricity and magnetism, which was elaborated on and brought together in 1861 in a set of four statements or laws by the Scottish physicist James Clerk Maxwell (1831–1879). These laws and their supporting equations completely define the relationship between electric and magnetic fields and play the same role in electromagnetism that Newton's laws do in gravity and motion. They can be summarized in this fashion:

1. An electric charge produces an electric field.

2. A magnetic field exists between the poles of a magnet.

3. Electric fields are produced by changing magnetic fields.

4. Magnetic fields are produced by both changing electric fields and by electric currents.

The first two principles explain *static* electric and magnetic fields—that is, without any currents or changing currents. Maxwell's most significant contribution was in the fourth principle. He recognized that magnetic fields are not only produced by electric currents but are also produced by changing electric fields. After he set forth and elaborated on the four laws, he realized that laws three and four mean that electric fields and magnetic fields cannot be separated because each produces the other. From this realization, he predicted the existence of *largely invisible energy waves*—what we now call electromagnetic radiation. That is, based fundamentally on the third and fourth equations, Maxwell predicted the existence of oscillating electromagnetic "fields" moving through space like ripples in a pond radiating from their source.

Electromagnetic waves possess three characteristics: wavelength, speed, and frequency. As seen in Figure 10-1, each wave consists of a series of crests and troughs. Wavelength is the distance between adjacent crests. Speed is measured by the movement of the crests. Frequency is a quantity that indicates how many crests pass a given point in one second (usually measured in units of hertz, named after the German physicist Heinrich Hertz).

Light is the narrow band of electromagnetic radiation that is visible to humans and animals. Maxwell calculated that the speed of all electromagnetic waves (including light) is 186,000 miles per second. This led him to the postulation that these waves exist at all frequencies and wavelengths throughout the universe (as shown in Figure 10-1). The wavelength depends on the frequency or rate that the field is oscillating—the higher the frequency, the shorter the wavelength. Thus the waves of electromagnetic radiation cover the entire spectrum of radiation, with only a small portion (with wavelengths between 16 and 32 millionths of an inch) visible to humans in the form of light. As shown in Figure 10-1, the electromagnetic spectrum includes everything from radio waves (with wavelengths of a few yards to thousands of miles), to microwaves (one-tenth of an inch to a foot) to x-rays and gamma-rays, the most energetic electromagnetic radiation that

FIGURE 10-1

The Electromagnetic Spectrum

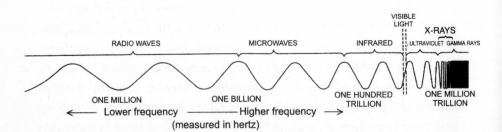

we can measure, which is produced in stars and as a result of the decay of some radioactive elements.

As we saw in Chapter 8, Einstein later relied on Maxwell's work in arriving at the special theory of relativity. He described Maxwell's theories as "the most profound and the most fruitful that physics has experienced since the time of Newton." Also, the knowledge of spectral lines enabled Edwin Hubble to detect the red shift of the galaxies moving away from one another, as discussed in Chapter 9.

Though Maxwell's equations led him to the prediction of the existence of electromagnetic radiation, it took another twenty-five years for that theory to be generally accepted—primarily because no such waves had ever been proven to exist or had ever been created in the laboratory until the Heinrich Hertz (1857–1894) did so in 1887, eight years after Maxwell's death. Hertz was the first person ever to broadcast and receive radio waves through ingenious experiments using the first primitive versions of transmitters, antennas, and detectors of radiofrequency electromagnetic radiation. This signaled the birth of radio communication, and of course, all of us now put the electromagnetic field to use in some practical way on a daily basis through the use of radio, TV, computers, cellular phones, and many other devices. As stated earlier, Hertz became the namesake for the unit of measurement for wave frequency.

Penzias and Wilson Accidentally Discover the Cosmic Background Radiation

Missing an Antenna

We must now pick up the story begun by Faraday, Maxwell, and Hertz fifty years later. As radio technology developed, scientists found that the sun's rays caused unwanted natural radio interference. In the late 1920s, Bell Telephone Laboratories wanted to set up a radiotelephone communication system across the Atlantic to supplement its cable system, and it needed to eliminate the interference. So Bell launched a research program to better understand the manner in which the sun causes the interference. The company constructed an array of antennas, receivers, and recording devices, which led to the discovery that radio emissions were coming not only from the sun but also from celestial bodies throughout the universe. The announcement of this discovery on April 27, 1933, gave birth to the modern science of radio astronomy. This represented a major advance in astronomy because millions of celestial bodies emit radio waves that can now be detected by radio telescopes even when optical telescopes cannot see such bodies. Astronomy is the oldest science, begun in China, Egypt, and Greece thousands of years ago. For the first time in its long history, astronomy no longer depended solely on optical observation. We will now see that the advent of radio astronomy led to a monumental discovery that even went beyond the detection of previously undetectable stars and galaxies.

Through the 1930s and 1940s, astrophysicists developed an understanding of the precise conditions that existed after the Big Bang. They realized that the enormously hot early universe gave rise to a thermal electromagnetic radiation field, including x-rays and gamma-rays—that is, radiation at the shorter wavelengths. As the universe cooled, the average temperature throughout it corresponded to longer and longer wavelengths on the spectrum. In 1948, physicists Ralph Alpher (1921–2007) and Robert Herman (1914–1997), working with the Russian-American physicist George Gamow (1904–1968), predicted that if the universe began in such a hot dense state, there should still be radiation or energy with an average temperature of five degrees above absolute zero (on the Kelvin scale) remaining from the Big Bang and spread thinly throughout the universe. They made this prediction in the paper mentioned in Chapter 9, "The Origin of Chemical Elements." Because there was no equipment to detect such residual radiation, their prediction went unnoticed for nearly two decades.

In 1965, German-American astrophysicist Arno Penzias (1933–) and American radio astronomer Robert W. Wilson (1936–), working as radio engineers at the Bell Telephone Laboratories in Holmdel, New Jersey, were calibrating a sensitive radio antenna that was being used to track the first Telstar communications satellite. They were attempting the find and eliminate the source of unwanted microwave "noise" that seemed to be registering with uniform intensity from all directions in space. The existence of these microwave signals (electromagnetic radiation with wavelengths between radio waves and infrared light) was very similar to the problem that Bell Laboratories faced in the 1920s, which led to the science of radio astronomy.

Although this new Bell antenna was designed to track satellites, it continued to pick up this pervasive radiation regardless of the Earth's rotation. This uniformity indicated to Penzias and Wilson that the radiation could not be coming from a particular celestial body or galaxy.

In the meantime, at nearby Princeton University, P. J. E. Peebles (1935–), a theoretical physicist, was working with a group on the radiation issue first addressed by Gamow, Alpher, and Herman in 1948. Peebles and the Princeton group had recalculated the Gamow-Alpher-Herman 1948 postulation entirely independently. Because they weren't aware of any equipment capable of detecting cosmic background radiation, they began to design an antenna intended to do so.

Back at the Bell facilities, Penzias and Wilson remained at a loss to explain the source of the static they found with the Telstar antenna. Coincidentally, in a telephone conversation with the well-known astronomer Bernard Burke (1928–) of MIT in Boston concerning an unrelated matter, Penzias happened to mention the problem of the unexplained microwave signals. Being aware of the Princeton work, Burke realized that Penzias and Wilson must have found the cosmic background radiation predicted by Peebles at Princeton, and he introduced the two groups. Penzias and Wilson and the Princeton physicists began working together, then simultaneously published papers in 1965 in the *Astrophysical Journal* that detailed their respective discoveries.

This marked a new era in research on the Big Bang and culminated in the information obtained by NASA's Cosmic Background Explorer (COBE) satellite in 1992. The equipment aboard COBE not only achieved the greatest accuracy to date on the temperature of microwave background radiation of photons, it also detected the minute variations (1 part in 100,000) in such radiation. This information has allowed cosmologists to re-create the approximate distribution of matter when the universe was only a million years old, which, in turn, has enabled them to figure out the events that led to the formation of the galaxies through the basic gravitational forces, as explained in Chapter 9.

Not only were hydrogen and helium forged in the inferno of the Big Bang, an invisible, steady, and pervasive field of microwave radiation—a fundamental form of energy moving through the cosmos at the speed of light—was caused by the incredible heat of that explosion. Georges Lemaître, the person who first proposed such an event in 1927 and lived to see it detected by radio astronomy, called this radiation "the vanished brilliance of the origin of the worlds." Theories and evidence accumulated in discrete and diverse quarters in the minds and laboratories of Maxwell, Hertz, Einstein, Planck, Hubble, Gamow, Peebles, Penzias, Wilson, and myriad others and synthesized into this inescapable conclusion: Fourteen billion years ago there was a titanic cosmic explosion which filled space with electromagnetic radiation that propagated throughout every corner of the universe. This radiation moved through the spectrum and left its weak remnants which can be discerned by radio telescopes today. From a distant and tiny spot in that "vast, cold, universal vacuum" we are awed and humbled as we have discovered "the vanished brilliance of the origin of the worlds."

Chapter Eleven
THE FATE OF OUR UNIVERSE

And the same hour was there a great earthquake, and the tenth part of the city fell, and . . . were slain of men seven thousand. . . . And the time of the dead that they should be judged. . . . And when the thousand years are expired, Satan shall be loosed out of his prison, and shall go out to deceive the nations which are in the four quarters of the earth. . . . Death and hell delivered up the dead which were in them; and they were judged every man according to their works.

—The Book of Revelation

We are leaving this Earth to find . . . a new dimension of truth and absolutism, far from the hypocrisies and oppression of this world.
—Member, Order of the Solar Temple (1994)

When will the universe end? What will it be like? Will anything come afterward? The belief in the apocalypse, followed by survival and peace for only the virtuous, is a myth that dates as far back as the fifteenth century BCE, and survives fully intact today. Each year various groups predict Armageddon with the same degree of certainty that Bishop James Ussher pinpointed Creation in his writings of the Old Testament, and each year catastrophes and natural disasters are said to be precursors to the final cataclysm. "Whereas the end of the world is approaching," began royal proclamations of tenth-century England, for many believed that Christ would reign for a millennium before the day of judgment, as noted in the book of Revelation, which is an entire body of apocalyptic prophecies. Indeed, the word *apocalypse* comes from a Greek word meaning "revelation." Today, we're reminded of the Branch Davidians in Waco, Texas, in 1993, and their predictions of chaos and destruction, and the groups that still stock their bomb shelters with provisions to survive the fire and earthquakes that will engulf the less worthy and the less resourceful.

In October 1994, fifty members of the Order of the Solar Temple in Quebec and the Order of the Solar Tradition in Switzerland committed suicide. These religious sects engaged in apocalyptic prophecy that focused on an impending environmental disaster that only a select few would survive. We continue to face our destiny and deal with elements and feelings of despair that have haunted humanity in every age.

The first recorded example of apocalyptic faith is in the Avesta, sacred scriptures of the Persian empire. The Avesta recounts the Iranian prophet Zoroaster's vision in which he saw the Spirit of Destruction (Angra Mainyu) confront the good god (Ahura Mazdā). Zoroaster lived sometime between 1200 and 1500 BCE, and predicted a vast transformation and judgment of all humanity arising out of this battle in which only the just would survive.

Apocalyptic predictions are also an integral part of Christianity and Judaism. In the Gospel according to Mark, Jesus overcomes the wrath of Satan to restore the communion between God and man, and in the book of Revelation Jesus appears as a Zoroastrian hero who defeats Satan at the battle of Armageddon. The Old Testament reflects apocalyptic movements of early Jewish sects in the destruction of Jerusalem and the Temple by the Babylonians and in those who sought to destroy Satan and plunge the sinners into eternal torment, while the righteous would be rewarded with eternal joy. In 2003, *The Psychology of Religion: An Empirical Approach* reviewed a sociological study carried out in an attempt to understand the mind-set of a group and the individuals involved in apocalyptic conspiracies. Titled "When Prophecy Fails: A Social and Psychological Study of a Modern Group That Predicted the Destruction of the World," this original study from 1956 has become a classic in the field of apocalyptic myths.

Common threads account for all apocalypse myths. We want to know about the end of the universe and our actions are believed to bring consequences. Most myths teach that there is reward for good, punishment for bad. That there is right and wrong. Morality must have a practical impact on our human actions and interactions or else there's less incentive to adhere to such moral principles. Myths reinforce this hope and fear—hope of eternal reward, and fear of the apocalypse, the cataclysmic end of the universe—and they attempt to add an element of meaning and urgency to life.

These ancient myths were fabricated out of fear and hope, yet they embody one element of truth and reality in our dynamic universe: *what came into existence must someday cease to exist.* People continue to fear and to be saddened by the prospect of humanity and the universe ending. Having examined the Big Bang and the formation of the universe, we'll now examine the most likely processes that could take place to cause the end of life on this Earth, the end of our solar system, and the end of the entire universe. These theories are based on observed and known scientific evidence and on the opinions of respected cosmologists and other scientists, though it's impossible to be certain which particular fate awaits us. Contrary to all apocalyptic teachings, it will happen without regard to any moral judgment and will spare no one.

Comets, Asteroids, and Climate Changes Threaten Life on Earth

Routine Collisions . . . the Fate of the Dinosaurs

The viability of life on Earth, but not the Earth itself, could fall victim to a huge comet or asteroid. As mentioned in Chapter 9, there are trillions of comets trapped by the gravitational fields of planets and the sun traveling within our solar system. There are also thousands of asteroids in the asteroid belt between Mars and Jupiter. The great mass of Jupiter sometimes captures comets, as in 1994 when the comet Shoemaker-Levy 9 crashed into the planet's southern hemisphere, vaporizing clouds and creating huge plumes in its atmosphere. It was the most cataclysmic collision in recorded history. Occasionally, Jupiter also flings asteroids toward Earth or past Earth toward the sun. It is estimated that there are about 10,000 comets and asteroids in orbits that intersect Earth's orbit and are large enough (at least a half mile in diameter) to create catastrophic consequences if they were to collide with our planet. Measured by geological time, collisions are fairly routine on Earth—a large comet or asteroid strikes the Earth on the *average* of every few million years, as was the case with the comet that hit the Earth 65 million years ago and is now widely believed to have caused the extinction of the dinosaurs. On February 15, 2013, a 10-ton meteor appeared in the skies over the Ural Region of Russia as a brilliant fireball traveling at more than 40,000 miles per hour. Even though the atmosphere absorbed most of its energy, the meteor burst at about 15 miles above ground with a force of more than 500 tons of dynamite. More than 1,000 people were injured and thousands of buildings were damaged across the region. This meteor was not detected before entering the atmosphere and was the largest object to have reached Earth since 1908.

Other possible threats to life on our fragile and tiny planet include the possibility of drastic alteration of the heat flow from the sun caused by interference from the giant clouds of celestial gas drifting throughout the solar system, slow ecological degradation, or gradual climatic change such as we are witnessing now in the form of rising global temperatures and consequent melting of glaciers and polar ice. Indeed, as of the year 2012, ice stored in Greenland and Antarctica has been measured to reveal an average loss of 344 billion tons of glacial ice per year over the past century. This has contributed significantly to a rise in sea levels worldwide.

The Sun's Fate Is Sealed As a Red Giant and White Dwarf

Humans Perishing in Advance

Looking at the larger picture, scientists have calculated that the entire solar system—including Earth—is certainly doomed 4 or 5 billion years down the road when our sun nears the end of its life. The sun is about 5 billion years old and can burn for another 5 billion years as it continues turning hydrogen into helium through fusion, the process that gives all stars heat and light. As the sun transforms all the hydrogen in its core to helium, gravitational energy will be released, causing the sun to expand because it will contain less mass—that is, less "self-gravity." Its surface will become somewhat cooler—to 6,700°F from its present temperature of 10,800°F—giving it a reddish hue. Ultimately, it will become a "red giant," a huge fireball with 200 times its present luminosity and about 30 times its present diameter, which will mark the beginning of the end for a star the size of our sun. Astronomers have viewed numerous red giants in the universe, such as Aldebaran, Betelgeuse, and Arcturus. Before the red giant phase, the sun's immense radiating surface will generate much greater heat, even though it will have less mass and lower surface temperature. Human beings and all other life on Earth will have perished long before the sun becomes a red giant because the Earth's oceans will boil away and the atmosphere will be chemically transformed by the tremendous heat. As the fireball increases in size, it will engulf the four innermost planets (Mercury, Venus, Earth, and Mars), one by one. Then, after billions more years of erratic and violent activity, our sun will eventually condense into a ball the size of a small planet and become an object known to astronomers as a "white dwarf," a star of incredible density and low luminosity. It will condense to a dwarf because it will no longer contain enough gas pressure to support itself against its own inward gravitational pull. Ultimately, the sun will cool, dim, and fade into the darkness of space. Indeed, every one of the trillions of stars throughout the universe is doomed.

Expansion Forever at an Accelerating Rate

Forces across the Cosmos

This leads us back to the initial question: *Will the entire universe end? If so, how?* We've looked at a few scenarios regarding the fate of life on Earth and the fate of the sun, the solar system, and the stars. *But what is the fate of the entire universe?* The amount of matter in the universe and the resulting force of its gravity is the central factor in answering these questions. Gravity is the weakest of nature's forces when examined at the atomic level but is the dominant force on the astronomical scale, having a range of billions of light-years and holding galaxies and clusters of galaxies together across the cosmos. If the rate of expansion of all the matter that exploded and is retreating from the Big Bang is fast enough, all such matter will be able to escape from the cumulative gravity of all the other material in the universe. Rockets, for example, are given sufficient "escape velocity" (about 25,000 miles per hour) to escape the Earth's gravity. The escape velocity that would allow an object to permanently leave the gravitational pull created by *the mass of the rest of the matter in the universe* is close to the speed of light. As mentioned in Chapter 9, it appears that the most distant galaxies in the universe are actually receding from the rest of the matter in the universe at this incredible speed, thus fast enough to escape, never to return, as discovered by recent Nobel recipients (discussed later in this chapter).

Will the universe ultimately lose the battle with *gravity*? In 1992 astrophysicists at Lawrence Berkeley Laboratory at the University of California at Berkeley discovered evidence that as much as 90 percent of the matter in the universe is so-called cold, dark matter in the form of massive wisps of gas spanning two-thirds of the universe (10 billion light-years or 59 billion trillion miles across). The matter was detected by measuring its microwave radiation with instruments on NASA's COBE satellite launched in 1989, as discussed in Chapter 10. Other "missing" mass is in the form of burned-out stars. The wisps of gas and these stars are invisible to the most powerful telescopes.

Another project dedicated to finding the answer to the ultimate fate of the universe is the collection of twenty-seven radio telescopes in Socorro, New Mexico, called the Very Large Array (VLA). The VLA detects the most distant objects in the universe, thus defining its "edge." Viewing those faint objects, we are literally looking into the past. That is, we see them as they existed billions of years ago because it has taken their light that long to reach us. Thus with the VLA and other instruments, astronomers measure the volume of matter and its recession speed. Most important, comparing the speed of recession of the distant outer objects (that is, their speed billions of years ago) with the *present* speed of recession of other objects has confirmed that the expansion is ongoing at increasing velocity and will continue for infinity.

Despite a theory that reigned for many years, the now-accepted view among cosmologists is that gravity is not slowing the expansion of the universe that was predicted to eventually fall back on itself. In other words, there will be no "Big Crunch" of a retracting cosmos. It now appears that just the opposite will occur as the universe expands at an increasing rate. That is, the expansion that began at the time of the Big Bang will continue forever. This was proven through publications in the *Astronomical Journal* in 1998 and in the *Astrophysical Journal* in 1999 by Professors Saul Perlmutter, Brian Schmidt, and Adam Riess, for which they received the Nobel Prize for physics in 2011. The discovery was so profound that the Nobel was awarded a brief thirteen years after the original publications, instead of after the average of over thirty years.

A large team of investigators made their initial observations in the early 1990s using the four-meter Blanco Telescope at the Cerro Tololo Inter-American Observatory in Chile. What they were looking for were supernovae, exploding stars that provide a measure of how far they are from Earth. Super-distant supernovae are receding from Earth so fast that the light they emanate is "stretched" into the red end of the electromagnetic spectrum. The "redshift" values were found in reference patches of sky just after a new moon and then those values were subtracted from the same sky imaged just before the next moon. Using algorithms designed by the Nobel laureates, the astronomer-physicists were then able to calculate how far the supernovae were from Earth and how quickly they reached their new distances from the Earth. Another key discovery made at the observatory was that the time it took for brightness of a supernova to fade away was proportional to its peak brightness. Thus once the ultimate brightness of a supernova was determined, the number of days it took for an exploding star to dim could be used to determine the distance from Earth and velocity of its movement away from the telescope. In this current scenario, the time it would take for the linear scale of the universe to expand to double its size would be approximately 11.4 billion years. Eventually, galaxies beyond our own will be so far into the redshift zone of the electromagnetic spectrum that it will become hard to detect them, and the distant universe will be in complete darkness.

During the final millions of years, some scientists predict that there will be a proliferation of black holes that will suck up all matter at an enormous rate until everything collapses into a single point of space and time with infinite temperature and density—perhaps to repeat the Big Bang all over again in a "re-creation" of the universe. For example, the black hole that forms the center of our Milky Way galaxy is named Sagittarius A (Sgr A). This "hole" is calculated to be 4 million times as massive as our sun but only thirty times as wide. Black holes are believed to evolve along with their galaxies, to exhibit strong gravitational fields, and to gather nearby matter by a process called accretion. Jets of material have been observed as emissions from black holes. Imaging the surface of a black hole has not yet been accomplished and would constitute the viewing of an "event horizon," the term for the point at which the gravitational pull of the black hole is so great that not even light can escape. Astronomers in 2012 combined radio telescopes from around the world to finally "see" this supermassive center of our galaxy. Black holes are difficult to image because they emit little light except for very weak radiation called Hawking radiation, but the combined telescopes that form the Event Horizon Telescope should be capable of imaging Sgr A within the next few years. Throughout the universe, there are also lighter "stellar-mass" black holes produced when massive stars collapse at the end of their lives. Fortunately, the U.S. National Science Foundation and observatories in several countries support this important scientific effort.

Another scenario on the fate of the universe has been proposed by a number of theoretical physicists who have picked up Einstein's work on the unified field theory. Like Einstein, they believe there exists a theory as yet undiscovered by anyone that will explain, in one unified set or series of principles and equations, the existence and properties of all matter and energy. This is now commonly called the "grand unified theory." One aspect of most of the grand unified theories now being developed is that *all protons in the universe will fall apart in several trillion trillion trillion years.* If all protons are destined to disintegrate on their own over the next several trillion trillion trillion years, as the theory predicts, every atom will also fall apart. Physicists from the University of California at Irvine, the University of Michigan, Brookhaven National Laboratory in Upton, New York, and other locations around the world are attempting to prove or disprove the theory in the hope of better understanding subatomic particles and tie them to all physical phenomena in the ultimate theory in physics. In 2010, the brilliance of Stephen Hawking appeared again in his book *The Grand Design.* He attempts to unify the forces of matter with those of gravity. Hawking explains that earlier attempts to understand the physics of unification were presented as "string theory," where elementary particles appear as vibrations along infinitesimal strings rather than as points. He also provides evidence for cosmic microwave radiation from the earliest moments after the Big Bang. Hawking's unification theory says that gravity accounts for negative energy, whereas the mass of stars clearly is positive. Thus if the overall energy of the universe is a sum of all the black holes, stars, and empty space, the result is zero. This, according to Hawking, unifies gravity with "quantum theory," according to which nuclear and radiation phenomena are explained through the assumption that energy occurs in discrete amounts known as quanta. One conclusion from his theory is that if the total energy of the universe is zero, it could have formed spontaneously from quantum fluctuations.

Whether or not Hawking's complex calculations and remarkable predictions prove to be true will likely consume the lives of physicists and astronomers for decades. But this is the sphere of influence where our future endeavors must lie, for as we move further into the twenty-first century, the Millennium Watch Institute, which tracks apocalyptic prophecies, has reported that about 1,000 organizations and individuals in the United States alone publish books and articles about global transformations that will lead to the apocalypse. Some warn of floods and earthquakes, while others predict war or famine. Lacking the long and rich history and tradition of these myths, the truth about our fate lies in understanding the physical principles that control asteroids, comets, the sun's fusion process, the cumulative gravity of the universe, and the theory of proton disintegration.

Looking back on the discoveries of Newton, Rutherford, Bohr, Einstein, Hubble, and the other scientists discussed in the these pages, we see that each successive step in acquiring scientific knowledge has given us a new perspective not only on the physical universe but also on the human race. As we've moved from the pre-Copernican center of a finite and changeless universe of 5,000 stars to the post-Hubble fringes of a galaxy of 300 billion stars amid billions of other galaxies, we keep testing the limits of our knowledge. Relative to the universe as we now know it, our entire world is smaller than a grain of sand among the other grains on all the world's beaches.

The universe began as an empty space filled with elementary particles, and it appears that it will expand ever more rapidly into infinity. Between this beginning and some undefined end, humanity has arisen, with its diverse and fascinating history and culture. As we continue to apply our reasoning ability to the understanding of the physical world, and until life is actually found elsewhere in the universe, we humans appear all the more incredible. We are more aware of the context in which we exist. We inhabit a brief interval of time with barren lifeless planets as our companions, all a result of the Big Bang. We exist in a remote corner of this galaxy in which the next closest star is *23 trillion miles away*. Though people have found comfort throughout history in the rich tradition of myth, there is greater comfort in what we now know, for it is rich in truth.

Now that we've explored the major discoveries in physics and astronomy, we will examine directly and in a similar depth this *matter that lives*—the unique and magnificent configurations and beings that we call life. We will explore the great mysteries of life and the intricate ties between the physical universe and the living universe that arose on Earth. We will see the relationship between the living material and the chemical elements of the periodic table. In Parts Five, Six and Seven, we will place life forms in their proper context amid the atoms of which they are constructed and the physical forces and environment that have molded and determined life's origins and destiny. We will systematically identify the conditions, steps, and developments that occurred over billions of years as the elements present in Earth's primordial soup first coalesced into life, diverged a million-fold, with some forms of life eventually climbing onto dry land and, in the case of one such form of life, evolving to the point where it is self-conscious and able to reason and analyze, able to preserve and communicate facts and concepts over generations by developing symbols we call letters, and by fantastic inventions able to fulfill its passion for understanding the most complex questions and to look back deeply into the fog of time and discern its own history.

PART FIVE

Evolution and the Principle of Natural Selection

In Focus

Of the seven greatest scientific discoveries, three stand out as completely changing our fundamental concepts about life and the universe: (1) gravity and the physics of the universe, (2) the Big Bang, and (3) Darwin's theory of evolution, which we will explore in the following pages. Evolution is far from a theory—it is a principle beyond question and beyond debate. Nevertheless, with their destructive and irrational agenda, twenty-first century creationists, straight out of the Dark Ages, seek to place emotion before reason and make evolution a debatable issue. The truth is, all forms of life share a common ancestor, and through a process that naturally "selects" the characteristics that allow it to survive and pass on the genes that code for those characteristics, each species, including humans, gradually has adapted and will continue to adapt to its changing environment—although, uniquely, humans are able to so profoundly alter their personal environments that natural selection within our species has been largely rendered inoperative.

In the 140 years since Darwin wrote *On the Origin of Species*, we've made astounding advances in understanding how evolution works, including twentieth and twenty-first-century fossil discoveries of *Homo sapiens'* predecessors and the precise mechanisms of evolution revealed in the cell and

DNA, which we will discuss in Parts Six and Seven. Also, a number of unexpected discoveries in other fields have filled in the picture. For example, in Chapter 15 we'll look at the role of plate tectonics in changing environmental conditions, in turn driving natural selection.

Charles Darwin's discovery approaches Newton's in significance and is similar in the sense that both men developed their propositions with minimal reliance on their predecessors. Their work was largely individual and incredibly original and comprehensive. Darwin has been described as "the Newton of biology," for Newton opened the door to our understanding of the universe, and Darwin opened the door to our understanding of life.

Chapter Twelve
THE ROCKS OF GENESIS

In the beginning God created the heaven and the earth. . . . And God created . . . every living creature that moveth. . . . And the Lord God formed man of the dust of the ground, and breathed into his nostrils the breath of life; and man became a living soul.
—The Book of Genesis

The belief that species were immutable productions was almost unavoidable as long as the history of the world was thought to be of short duration. . . . The chief cause of our . . . unwillingness to admit that one species has given birth to clear and distinct species is that we are always slow in admitting great changes of which we do not see steps.
—Charles Darwin, *On the Origin of Species* (1859)

We now turn from the telescope and distant heavens and begin to look at the origin and nature of life on this Earth. As dangerous and controversial as Galileo's and Newton's discoveries were thought to be, those who would dare challenge fundamental beliefs about life itself, particularly about humankind, would find themselves in the midst of even greater danger and controversy. That should have been expected—the origin of life is a much more personal and more emotional subject than the Earth's location or the nature of gravity. But unfazed by their many detractors, a new group of scientists emerged in the nineteenth century and began a revolution in geology and biology that resulted in an understanding of the age of the Earth and the nature and origin of living things. This new understanding was so profound it removed humankind's childlike innocence forever.

Spontaneous Generation of Life Is Disproven but There Is No Reason to Question Creation

In God's Image

Until the Renaissance, reproduction was believed to be a result of a supernatural event, permanently beyond the descriptive powers of science, the same as creation of life in Genesis. Reproduction of the simpler forms of life was thought to occur spontaneously out of nonliving matter, such as the sudden appearance of maggots from meat, and the generation of worms from trees, beetles from dung, and mice from garbage. This set of beliefs was called "spontaneous generation." Another notion common in academic circles of the seventeenth century was the preformation theory, which held that the sperm or the egg (depending on which version one supported) contained microscopic preformed beings created for all future generations at the time of Adam and Eve, awaiting fertilization to trigger a mysterious growth mechanism. The Italian physician and biologist Marcello Malpighi (1628–1694) popularized this theory, though his decades of work as the founder of microscopic anatomy and other real scientific advances in embryology, anatomy, and pathology far outweigh any negative impact on advances in biology resulting from his belief in the preformation theory.

The British physiologist William Harvey (1578–1657) was first to demonstrate that spontaneous generation was impossible because every animal comes from an egg, and the Italian biologist Francesco Redi (1626–1697) conducted experiments showing that the presence of maggots in putrefying meat is not caused spontaneously, but from microscopic eggs laid on the meat by flies. Through his decades of research on bacteria, protozoa, and other simple forms of microscopic animal life, the famous Dutch lens maker Antonie van Leeuwenhoek (1632–1723), who made great improvements to the microscope, contributed much to dispelling the idea of spontaneous generation. However, he also promoted the preformation theory. In the late 1700s, the Italian priest and physiologist Lozzaro Spallanzani (1729–1799) conducted a series of experiments proving that actual contact between egg and sperm is essential for the development of new animals.

Biological explanations for reproduction were gradually recognized and accepted during the 1700s. Yet the entire field of biology—the understanding of all living things—remained in its infant stages. Until about 1830, the common and accepted view of the world was that we lived on a stable and unchanging planet populated by immutable, unchanging species, just as God created them, and that He created Earth itself to be a home for man. All plants and animals were believed to be eternal in their present form—an understandable view, for the theory of a changing universe with a beginning and end was not developed until the twentieth century, with the discovery of the Big Bang and recession of the galaxies, as we saw in Part Four. Aristotle's crystalline spheres had been shattered well before 1830, and people had accepted the idea of "Antipodes" (a southern hemisphere with places geographically diametrically opposed to the centers of Western civilization). However, merely because the Earth rotated and revolved, and simply because Newtonian physics applied equally to New Zealand and Great Britain, did not mean Earth itself or the life on it had physically changed over time.

Except for the Earth's location in the solar system, the Bible was still taken literally. The universe was created in six days. As mentioned in Chapter 9, Archbishop James Ussher of Ireland calculated the year of creation to be 4004 BCE and the *moment* of creation to be 2:30 p.m., Sunday, October 23. This belief made human existence clear and comprehensible. People drew comfort from their feeling that God had not cruelly withheld our own past from us—that we hadn't missed anything since the beginning of the universe because there was no universe, no history, before the Bible. The word *prehistoric* did not enter European vocabularies until the mid-nineteenth century. Using 50 years as an average lifetime, 5,800 years is only 116 consecutive lifetimes. Thus humankind had emerged from an unknown darkness 116 lifetimes ago, much like emerging from the womb. We believed ourselves to be children of God, innocent and protected, and we had no basis to doubt or question that belief. The familiar idea of birth was adapted to the concept of creation. God created Adam and gave birth to humankind, as depicted in Michelangelo's painting on the ceiling of the Sistine Chapel in Rome. The art and literature of Western civilization embedded this image deep within our culture. It would take another revolution to wrench that image from our minds.

Categorizing and Comparing Living Things Leads to the Question of Whether Life Forms Are Permanently Fixed

The Sixth Day of Creation

Aristotle had taught that everything in nature seeks perfection, reaches for perfection, with some achieving (like humans), and others failing (like all lesser animals), but that living things do not evolve in the biological sense. His elaborate "Scale of Being" philosophy was a living ladder that placed minerals at the bottom and man at the top, with nothing changing. Species were eternally fixed signposts identifying the levels of perfection or imperfection, having nothing to do with their anatomy or actual evolutionary relationship. It was this concept, coupled with Genesis, Noah's Ark, biblically described humans, and the power of religion that directed rational thought away from any idea of evolution and allowed Aristotle's fixed-species concept to outlive his fixed-Earth concept by two centuries.

In 1691, the popular naturalist John Ray (1627–1705) wrote *The Wisdom of God Manifested in the Works of Creation,* which described a pattern of living things based on their structural characteristics and internal anatomy. This work and others perpetuated and refined Aristotle's concept by logically organizing, identifying, and categorizing the living things known at that time and giving them a nomenclature, which rigidly fixed the immutability concept. At the same time, however, it had the paradoxical effect of laying the groundwork for comparative anatomy, which later became an important part of evolutionary theory. That is, evolutionary theory first needed this clear picture of anatomical similarities and differences of species on which to build.

The most significant figure in the trend to classify, describe, and compare was Carl Linnaeus (1707–1778) of Sweden. The great modern anthropologist and philosopher Loren Eiseley described Linnaeus's love of the incredible variety of life as a "poetic hunger of the mind to experience every leaf, flower, and bird. . . . He was the naming genius par excellence." In *Systema Naturae,* first published in 1735, Linnaeus created a clear and efficient method of naming all animals and plants. Each was identified by two names. The first was generic to identify a group (or genus) of visibly related creatures, such as all doglike forms. The second referred to a more specific group, a species, such as wolves. Thus the wolf is *Canis lupus*. In addition to the genus and species reflected in the name, the Linnaean system (as modified and refined by modern biology) has five increasingly more specific levels, each with additional categories. Thus the seven levels of all living things (kingdom, phylum, class, order, family, genus, and species), beginning with the largest and most general, consist of first a kingdom. Biologists from the first quarter of the twentieth century recognized two major groups of life: prokaryotes and eukaryotes. The first group lacks a complete nucleus (see Figures 16-1 on page 267 and 16-4 on page 278), the central cellular organelle that contains DNA and the mechanisms controlling cell growth. These most primitive prokaryotic forms are now divided into two groups, the Eubacteria and Archaea, each within their own kingdom. The former are "true" bacteria (see Figure 16-4), and the latter are bacteria-like single cells but exhibit a great deal of genetic diversity from the Eubacteria. The Archaea actually have more genetic similarities to the eukaryotes than to the true bacteria, indicating an independent evolutionary pathway. The Archaea have similar size and appearance to many bacteria but are often recognized as "extremophiles"— that is, organisms that flourish in extreme environments such as hot springs, salt lakes and oceans, marshlands, and the colons of humans and other animals. The fascinating organisms identified in this kingdom Archaea and their roles in the natural processes of carbon cycles and food digestion are currently under intense investigation. Several species of Archaea produce the spectacular orange-colored rings around the boiling sulfur springs of Yellowstone National Park. The discoverer of this "third domain of life" was the Nobel laureate and iconoclast Carl Woese (1928–2012). In contrast, members of the kingdom Eubacteria have been studied extensively by generations of scientists because they play such a profound role in causing disease and in mediating normal processes required by animals and plants (see Chapter 16).

All organisms in the following familiar four kingdoms are eukaryotes (with a complete nucleus): Protists (single-celled plants or animals), Plants, Fungi, and Animals. Each of the kingdoms is broken down into increasingly smaller groups of organisms until the final genus and species as originally designed by Linneaus. These groups are:

Phylum: For example, there are twenty phyla in the animal kingdom. One is Chordata—all animals with head-to-tail nerve cords (fish, snakes, bats, humans, etc.)

Class: For example, one class of chordates is Mammalia, which are animals that have mammary glands to nurse their young, fur or hair, and other specific characteristics (bats, whales, humans, etc.)

Order: For example, one order of mammals is Primates, which possess all of the characteristics described for mammals and chordates and also have five fingers and toes, nails instead of claws, eyes that face forward, hands that can grasp because of an opposable thumb, and a relatively large brain (tarsiers, baboons, chimpanzees, humans, etc.). As you can see from this classification, humans share the same kingdom, phylum, class, and order with apes.

Family: For example, one family of primates is Hominidae (which are also mammals and chordates), characterized by their ability to walk upright on two legs.

Genus: For example, one genus of hominid is *Homo* (from the Latin, meaning "human," not the Greek, meaning "same")—characterized by larger brain size and a head shaped like modern human.

Species: Of the genus *Homo* and family Hominidae, *Homo sapiens* (*sapiens* meaning "wise") is the only extant (still living) species. There are a number of extinct species of hominids. Three well known examples are *Homo neanderthalensis*, *Homo habilis*, and *Homo erectus* (discussed in Chapter 14).

As another example, the Blackburnian warbler (*Setophaga fusca*) is one species that shares the genus *Dendroica* with 27 other species, and the family Parulidae with 125 more species of little warblers. In modern terms, a species means all the members of a group that are genetically capable of interbreeding and producing fertile offspring.

Linnaeus's system was extremely useful and timely during the eighteenth and nineteenth centuries, with the increasing number of sea voyages that resulted in the rapid discovery of thousands of new species of plants and animals in the New World, Australia, Africa, and the Far East. Linnaeus rose to fame as a result of his mastery of the living world and the order he imposed on it. Largely because of his development of this system, naturalists were given official posts on board overseas voyages, such as the one that was to set Charles Darwin on the course of evolutionary theory.

In the early editions of *Systema*, Linnaeus maintained the idea of the fixity of species—that modern species were as fixed as on the sixth day of creation. His system of class and gradation made it more clear than ever that each living thing possessed a specific form. However, as Linnaeus continued his lifelong work, the paradox began to surface as he found it increasingly difficult to cling to his original thesis of the immutability of species. Recognizing the results of crossbreeding and inbreeding farm animals and the existence of small inherited changes called mutations, he became confused as to "whether all these species are the children of time, or whether the Creator from the very beginning of the world had restricted this course of development to a definite number of species." With this unanswered question in mind, he removed from later editions of *Systema* his assertion that all species are fixed and no new ones can arise. So Linnaeus's system was not only a basis for comparative anatomy of living things but one that could reflect change in those living things.

Modern Geology Is Stillborn as Creationism Persists

Time Is of the Essence . . . Mammoth Elephants and Saber-Toothed Cats

As early as 560 BCE the Greek philosopher Xenophanes of Colophon (570–475 BCE) found seashells embedded in the rock strata high in the mountains of Greece. He believed that the seashells became embedded there during a catastrophic flood. In the mid-1600s, the French naturalist Isaac de la Peyrère published a book about some strangely shaped stones that had been collected in the French countryside, and suggested that the stones had been shaped by primitive men who lived before the time of Adam. De la Peyrère's life was threatened and his book was burned publicly in 1655.

Then, in 1749, the French naturalist Georges-Louis Leclerc, Comte de

Buffon (1707–1788), published the first of his thirty-six-volume work on natural history. This monumental study reflected Buffon's revolutionary ideas in geology and biology and was the first work to treat those subjects scientifically. He concluded that some animal life had become extinct and discreetly suggested that animals might have undergone some type of evolutionary change. He also stated that some mammals might have *common* ancestors that were physically different from those mammals of his day. Scientists and theologians were critical of Buffon's ideas of evolution, his concept of geological history in stages, and his suggestion of a time scale of up to 35,000 years to account for such stratification and the history of life on Earth. Just as the Catholic Church had forced Galileo to recant in 1633, the theology committee at the University of Paris forced Buffon to retract in writing certain passages to avoid censure: "I abandon whatever in my book concerns the formation of the Earth," said Buffon, "and all that might be contrary to the narration of Moses."

In 1771, Johann Friedrich Esper (1732–1781) found human bones in a cave in Germany along with the skeleton of an extinct bear. Near Hoxne, England, in 1790, the British archaeologist John Frere (1740–1807) found some Stone Age flint implements among the bones of other animals that no longer existed on Earth. In other locations in Europe, human skulls were found with remains like these and with numerous other types of fossil discoveries—all forming an image not found in the Bible, an image of an Earth that was once inhabited by mammoth elephants, woolly rhinos, saber-toothed cats, and many other species, all now extinct.

However, apart from the need for fossil evidence on the *biological* aspects of such a revolutionary theory as evolution, there first needed to be evidence of sufficient *time* for it to occur. Buffon had speculated on this point, but the crucial evidence was locked in the Earth itself, waiting to be liberated through careful observation and analysis by those who were to develop a completely new field of science. After years of such observation and research, James Hutton (1726–1797) of Scotland presented the first well-documented analysis on the age of the Earth to the Royal Society of Edinburgh in 1785, in a paper titled "Theory of the Earth." Hutton described a compelling picture: *soils form by the weathering of rocks; tides and the pounding of waves erode the coast; layers of sediment accumulate; and the general cycles of sedimentation, uplift of hills and mountains, and erosion can be seen everywhere.*

With this paper, Hutton created a new scientific discipline (geology) from his own unique observations and became the father of modern geology. He originated the fundamental principle on which geology is now based, called uniformitarianism or uniformity, which holds that the rocks and other inorganic materials of the Earth are formed and modified by a continuous and generally uniform series of natural phenomena, such as rain, wind, tides, and gradual shifts in the Earth's crust. The impact of catastrophes such as floods and earthquakes was recognized by Hutton, but as exceptions to the basic geologic processes of uniformitarianism. "The present is the key to the past," said Hutton, meaning that all the conditions responsible for observed (past) geological phenomena are the same conditions that exist today. That is, past phenomena can be fully explained by current processes *plus time.*

Ten years after his original presentation to the Royal Society, Hutton compiled a more comprehensive analysis in a two-volume work, *Theory of the Earth.* He squarely proposed that the Earth was at least hundreds of thousands of years old. "James Hutton brooding over a little Scottish brook that carried sediment down to the sea," said Loren Eiseley, "felt the weight of the solid continent slide uneasily beneath his feet and cities and empires flow away as insubstantially as a summer cloud. . . . He discovered an intangible thing against which the human mind had long armored itself . . . time." Hutton opened this expanded window of time, allowing others the opportunity to place the growing number of fossil discoveries in their proper context.

Despite the force and persuasiveness of Hutton's lectures and writings, many scientists and others challenged uniformitarianism. They would not accept any concept that conflicted with the popularly accepted view called catastrophism, which held that geologic formations were caused by sudden catastrophic upheavals and events, such as Noah's flood, and that the Earth was created by God about 5,800 years earlier to be a home for humankind. Hutton was branded a heretic by many, but to a growing number of scientists it became apparent that catastrophism would have to be abandoned as a viable principle or modified to incorporate Hutton's theories, while somehow preserving the tenets of Christian theology.

In 1796, the French zoologist and statesman Georges Cuvier (1769–1832) began to accumulate his own evidence of the Earth's history. While a professor of natural history at the College de France, Cuvier discovered woolly mammoth elephant bones near Paris, giant salamanders, flying reptiles, and other extinct species. His fossils indicated a correlation between the rock strata and the development or changes in extinct species—that is, the deeper the strata, the less similar to animals now living, as compared to the fossils in the more recent strata. With the help of mineralogists, Cuvier tried to figure out how the bones of these fascinating extinct creatures had become entombed in the ground. He more fully developed the new scientific discipline of comparative anatomy (building on Linnaeus's foundation) through his brilliant writings, such as *Lessons on Comparative Anatomy* (1800), in which he stated his principle of the "correlation of parts," and *Researches on the Bones of Fossil Vertebrates* (1812).

It would seem that Cuvier had no choice but to confirm Hutton's theories, as the Earth revealed its ancient story told by layer upon layer of seashells, bones, different types of rock, the evidence of lakes and seas, and the recession of water followed by a new marine era. But as a devoutly religious man Cuvier saw his task to be reconciling the evidence of these phases of geological events with an Earth less than 6,000 years old, placing him in the difficult role of becoming the leading spokesman for catastrophism. He concluded that the popular concept of catastrophism was right and that James Hutton was wrong. A series of vast floods, said Cuvier, had occurred that accounted for the sedimentation and the extinction of species. He expounded on his version of catastrophism beginning in 1812, and wrote supporting treatises in 1815, with *Theory of the Earth,* and in 1825, *Discourse on the Revolutions of the Globe,* despite the increasing support for Hutton in a scattering of new writings, such as *The Stratigraphical System of Organized Fossils,* published in 1817 by William Smith (1769–1839). Cuvier maintained that the species spared from the sudden land upheavals and floods repopulated the Earth. He believed new species weren't actually new, but were from unexplored parts of the world. Cuvier's unfortunate legacy was to create a model for twentieth-century creationists.

Without a realistic concept of geologic time, catastrophism remained the popular view, preserving what were to be Aristotle's last gasps in biology, and perpetuating the divine origin and immutability of species.

Geologic Time Becomes a Reality, but Evolutionary Theory Remains Weak and Unsupported

In the Image of an Ape

The strongest support for a theory of evolution before the nineteenth century was developed through the research and writings of Erasmus Darwin (1731–1802) (Charles Darwin's grandfather), and Jean-Baptiste Lamarck (1744–1829). Erasmus Darwin was a prominent doctor in England and was also known as an innovative and radical thinker. His version of evolutionary theory held that species modify themselves by adapting to their environment through a conscious effort of some kind, which came to be known as the doctrine of acquired characteristics.

Jean-Baptiste Lamarck, a contemporary of Erasmus Darwin, was the better known of the two, and maintained a similar evolutionary philosophy. Lamarck was a French botanist and biologist, and made great contributions to the proper classifications of plant and animal life (class, order, genera, etc). In his 1809 publication, *Zoological Philosophy*, he set forth two principles that he believed explained the vast array and development of life forms: organs improve with use and are weakened by disuse, and such changes are preserved in animals and passed on to their offspring. So, for example, he believed that giraffes have long necks because they stretch to reach leaves in trees. This was sort of a "reverse Darwinism" in that the creature controls its own destiny, rather than the environment producing the conditions that influence the survival of creatures that are best adapted to them. It was conjecture, not based on any scientific approach or proof.

Lamarck's version of the doctrine of acquired characteristics was a combination of his own view of God's plan and Aristotle's Scale of Being. Lamarck wrote of life ascending toward higher levels to perfect the creation. Although he recognized that changes in the physical environment cause changes in the needs of living things, he blamed these environmental factors for *interfering* with all animals' abilities to achieve perfection in accordance with God's ultimate purpose. Like other eighteenth-century evolutionists, he viewed variations as being restricted within limits. For example, a gray pigeon and white pigeon might share a common ancestor from many generations before, but the dog, wolf, and bear surely could not have the same ancestor as the pigeon, no matter how long ago one could look back. Lamarck's general views were interesting and controversial and won support from some biologists in the early part of the nineteenth century. Yet his inability and refusal to support his theories with specific data led to his eventual ostracism. He died blind and impoverished.

So, by the 1820s no one had yet seen the full view through the window opened by Hutton. Geology and biology were not yet linked in any cohesive manner: Linnaeus had created the system of categorizing living things into seven levels; we had begun to view cells with the microscope and ask about the developmental stages of living things; and Hutton had made a strong case for uniformitarianism, but Cuvier's catastrophism blinded people from a view that allowed for the necessary geologic time. Buffon, Erasmus Darwin, and Lamarck were known in intellectual circles as suggesting that animals might not be as immutable as once thought, but the process and driving force behind evolution remained vague and the subject of conjecture and speculation. The suggestions of evolution and an ancient Earth were still faint voices drowned out by the popular views of fixity of species and catastrophism.

In the wake of Cuvier's popularity in Europe in the early 1800s, an English geologist named Charles Lyell (1797–1875) examined rock formations and fossils in greater detail than had ever been done before. He published *Principles of Geology* in 1830, in which he adopted Hutton's views and more convincingly demonstrated that despite Cuvier's pronouncements, catastrophism simply didn't make sense. Even though catastrophes occur from time to time, the natural forces of wind, water, and shifting earth have generally operated slowly and uniformly.

Lyell supported his principles with an unprecedented number of examples that he obtained through his extensive travels around the world. Faced with this overwhelming evidence, people finally began to accept the concept of a slow uniform geological process. For the first time in history it appeared the Earth might even be *millions* of years old. Indeed, perhaps the Earth had changed. Perhaps inorganic formations of rocks, beaches, and mountains had eroded and developed entirely new configurations. In geology, the tide began to turn in favor of Hutton, Lyell, and the young naturalists who were not mired in the past.

Nevertheless, even if the Earth was older than originally thought and had undergone change as described by Lyell, it remained unacceptable—scientifically, philosophically, and in particular religiously—to suggest that *life itself* had somehow evolved. Therefore, in 1830, when P. C. Schmerling found stone artifacts and two human skulls together with extinct rhinoceroses and mammoths in Belgium, the proximity was explained away as coincidence or fraud. In 1838, the French archaeologist and writer Jacques Boucher de Perthes (1788–1868) found a large number of flint hand axes and other stone objects near Abbeville in northern France that appeared to be ancient and hand crafted. Boucher de Perthes presented his finds to the scientific community with his theory that the hand axes were made by primitive men who existed before the deluge described in Genesis. Despite this and other mounting evidence, Boucher de Perthes's theories and writings were ignored, much as Isaac de la Peyrère's similar discoveries were rejected in the 1600s, because without radioactive dating techniques or the ability to chemically analyze these materials, there was no scientific method of proving their age.

Thus although Lyell's geologic principles continued to gain acceptance and the door slowly opened to question the *age* of Earth and its living things, this didn't connote evolution. It meant only that the immutable life forms created by God might have been in existence for longer than 5,800 years. It was one thing to accept a changing Earth and an expanded view of biological time, but the idea of changing life forms remained outrageous and the province of a few fanciful philosophers. The view stated by seventeenth-century British naturalist and botanist John Ray endured: "Whatever may be said for ye Antiquity of the Earth itself and bodies lodged in it, ye race of mankind is new."

Amid growing evidence of primitive humans; extinct species with human artifacts; the complete lack of fossils of existing species; and the work of Buffon, Lamarck, and Erasmus Darwin, the question persisted: *Do living things evolve?* If the inorganic world changed over time, as Hutton and Lyell had shown, is it possible the organic world also evolved over time? Why the extinct animals? The human bones in caves? Was man once as primitive as the tools and bones suggest? *Was man ever less than man?* Questions of explosive proportions.

As long as species were fixed in their present form and the Earth's changes could be attributed to sudden and cataclysmic causes, such as Noah's flood, there was peace between the theologians and scientists. That had been the state of affairs during the first few decades of the nineteenth century—an unspoken truce, for the geologists and biologists had gone as far as possible to reconcile their observations with the scriptures. But Hutton's and Lyell's changing Earth view, coupled with attempts at evolutionary theory, marked the beginning of the end of this uneasy compromise.

Against God's word, Copernicus, Kepler, Galileo, and Newton had taken humankind out of the center of the universe. Now another heretic was about to peer through the window opened by Hutton and Lyell and describe an unimaginable vista. In describing his view, he would offend humanity further by showing that Michelangelo's painting on the ceiling of the Sistine Chapel in Rome must be totally redesigned, for humans were *not* created in the image and likeness of God, and the fingertip endowing Adam with existence was not God's at all. *It was the fingertip of an ape.*

Chapter Thirteen
THE INDELIBLE STAMP
OF OUR ORIGIN

The great synthesizer who alters the outlook of a generation, who sud-denly produces a kaleidoscopic change in our vision of the world, is apt to be the most envied, feared, and hated man among his contemporaries.
— Loren Eiseley, *Darwin's Century* (1958)

Indeed, the boldest and most controversial revelation of the nineteenth century, and perhaps all of history, was the theory of evolution by natural selection developed by Charles Robert Darwin (1809–1882). It fired the emotions of Darwin's generation and each one that followed—those who hailed his genius, and those who vowed to burn him at the stake.

Darwin Sets Sail on the *Beagle*

Finches and Ostriches Offer a Clue

Darwin's mother died when he was eight, and his eldest sister, Caroline, raised him. He attended exclusive private schools as a boy but was not a good student, caring "for nothing but shooting, dogs and rat-catching," according to his father, a wealthy doctor. Reminiscent of Einstein's father's comment, the elder Darwin once told his son, "You will be a disgrace to yourself and your family." After dropping out of medical school at Edinburgh University, Darwin graduated as a theology student from Christ's College at the University of Cambridge, England, class of 1831, and intended to have a life of service in the Church of England. But there was no hurry about the ordination because there was usually a wait of a year or two before a deaconship or other position was available, and Darwin's father agreed to support him in the interim.

He'd done well in his naturalist classes in college, and considered taking a "geologizing" trip or a voyage on a merchant vessel to another part of the world to expand his horizons before making the lifelong commitment to the clergy. He admired his botany professor John Stevens Henslow (1796–1861), who'd made botany and biology popular subjects at the university and with whom Darwin had become friends, but he had never considered basing his life or profession on those subjects. However, in late August 1831, Darwin received a letter from a government official named George Peacock that led to a change in Darwin's views on the importance of biology, and would ultimately change the entire course of history:

My Dear Sir,

Captain FitzRoy . . . sails at the end of September in a ship to survey . . . the S. Coast of Terra del Fuego, afterwards to visit the South Sea Islands & to return by the Indian Archipelago to England. The expedition is entirely for scientific purposes & the ship will generally wait your leisure for researches in natural history. . . . The Admiralty are not disposed to give a salary, though they will furnish you with an official appointment. . . . If a salary should be required however I am inclined to think it would be granted.

The recommendation for Darwin's appointment as official naturalist on the British navy surveying ship H.M.S. *Beagle* came from Professor Henslow. It was a unique opportunity—a voyage around the world collecting and identifying every example of living plant and animal encountered. That was the extent of botany and biology as it existed at that time—categorizing living things and fitting them into the system established by Carl Linnaeus. The trip was to last no longer than two years, and then Darwin planned to return to the church.

He was selected despite being the *third* choice as naturalist on the *Beagle*. Leonard Jenyns (Henslow's brother-in-law) had been the first to be offered the position. Professor Henslow himself was the second. Jenyns declined because of his commitment to his church congregations, and Henslow declined because of his wife's objections. Darwin came close to being left on shore because of a personality conflict and differing political philosophy with the ship's temperamental Captain FitzRoy. Also, Darwin's father initially objected to the undertaking but eventually relented after lobbying by Darwin's uncle Josiah Wedgwood, the famous English pottery designer and manufacturer.

After all this and a series of delays by bad weather, the *Beagle* weighed anchor at 11:00 a.m., December 27, 1831, manned by a small crew. Charles Darwin was crammed into his living quarters in the poop cabin with barely enough headroom to straighten his six-foot frame, four inches on either side of him when upright, five feet of hammock, and room for few possessions. The ship took Darwin and the crew from Plymouth, England, to the Canary Islands, down the east coast of South America, up the west coast to the Galapagos Islands, then to New Zealand, Australia, Tasmania, Madagascar, around the Cape of Good Hope, the Ascension Islands, back to Brazil, and finally home.

His singular purpose during the voyage was to collect samples of rocks, fossils, sea animals, plant life, and animals on shore near the harbor towns where they stopped during their travels. His responsibility as naturalist included proper preparation and preservation of the specimens, packaging and shipping them back to Professor Henslow for observation, categorization, and analysis. As Darwin carefully observed and examined all these life forms, extinct and living, and the surrounding geologic formations, he began to develop an insight into the relationship between that changing geology and the life it sustains. He was particularly influenced by Lyell's *Principles of Geology*, a going-away gift from Henslow, and saw firsthand the cycles Lyell wrote about: the wear and tear of waves, the fall of forest and rock, gradual rising of land, and the erosion caused by rivers.

Contrary to common belief, Darwin did not develop his theory of evolution during the thirty-six days he spent on the Galapagos Islands, while the *Beagle* charted its shore during the fall of 1835. But the Galapagos did provide an essential piece of the puzzle by raising questions about the great variations he saw in the indigenous animal life there. Darwin studied the finches he caught on two different islands and found they had differently

FIGURE 13-1

The Voyage of the *Beagle*, 1831–1836

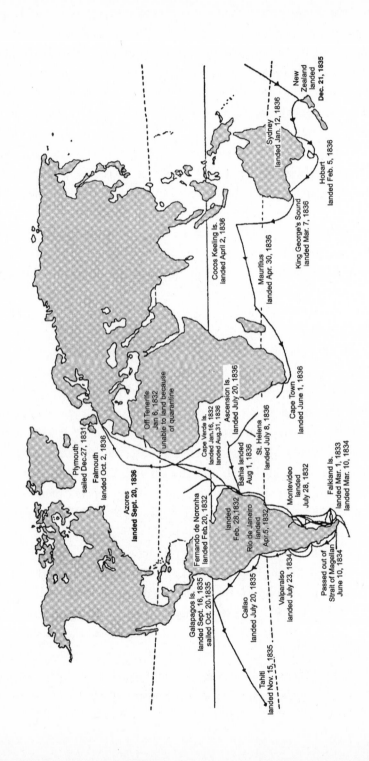

shaped beaks (Figure 13-2A and B). Similarly, he observed that shells of the tortoises of the various islands differed in shape, color, and thickness. He also noted differences in the tortoises' size and the length of their necks and legs. Distinct physical characteristics of these and several other animal species on the sixteen islands prompted Darwin to speculate about what causes the differences. Had each of the islands developed its own species? His experience in the Galapagos also led him to broader questions: Why are animals similar in places that are so very different? Why is the South African ostrich so much like the South American rhea? But then, why aren't the *fossil* animals found in the Galapagos identical to the *living* forms there? The voyage gave rise to these and other questions and to doubts about the literal truth of Genesis. By the end of this five-year awakening, Darwin began to scrutinize every specimen he encountered and to establish a revolutionary new paradigm for living things.

FIGURE 13-2
A. Darwin's Ground Finch

A. R. Brody

The beaks of Darwin's finches have multiple shapes and sizes that have adapted to local conditions and available food. The ground finch seen here has evolved a beak to crush large tough seeds. Compare to the beak of the finch seen in panel B.

B. Darwin's Tree Finch

A. R. Brody

The tree finch has a small sharp beak for picking softer seeds off bushes and trees.

The voyage that was supposed to take two years ended up taking five because of bad weather, mechanical problems, the need to resurvey where errors had been made, and all the unexpected events of a trip on the high seas in a small vessel in the early 1800s. By the time the *Beagle* docked at Falmouth, England, on October 2, 1836, Darwin had already gained some fame in England for the vast collection of specimens he'd gathered and sent to Henslow during the voyage. Soon after his return, he began presenting a series of papers before various prestigious scientific groups such as the London Geological Society on subjects ranging from fossil bones he'd dug out of the cliffs in South America to his theory of how corals are formed. Fascinated by the immense variety of species, their interrelationship, and the questions about their history, Darwin abandoned his earlier course as a man of the cloth and set a new one in nature.

The Theory of Evolution Is Born

One Common Ancestor

In mid-July 1837, Darwin began writing in his journal on the subject that had begun to take form on the Galapagos Islands: *the transmutation of species*. He usually avoided using the word *evolution* and preferred *transmutation* because he wanted to emphasize that species of plants and animals *change* in response to their environment but might remain *that* species. He later described how he came to address the issue:

> When on board H.M.S. *Beagle*, as naturalist, I was much struck with certain facts in the distribution of the organic beings inhabiting South America, and in the geological relations of the present to the past inhabitants of that continent. These facts . . . seemed to throw some light on the origin of species. . . . On my return home, it occurred to me, in 1837, that something might perhaps be made out on this question by patiently accumulating and reflecting on all sorts of facts which could possibly have any bearing on it.

At about the same time, Darwin happened to be reading a 1798 essay, "Malthus on Population," by the famous English economist and demographer Thomas Malthus (1766–1834), who maintained that human population and demand will always exceed the production of food and other needs. Darwin described Malthus's views in this way:

> There is a constant tendency in all animated life to increase beyond the nourishment prepared for it. . . . Nature has scattered the seeds of life abroad with the most profuse and liberal hand; but has been comparatively sparing in the room and the nourishment necessary to rear them. . . . Population has this constant tendency to increase beyond the means of subsistence . . .

In the course of reading Malthus in October 1838, he experienced a great revelation:

Being well prepared to appreciate the struggle for existence which every-where goes on . . . it at once struck me that under these circumstances favorable variations would tend to be preserved, and unfavorable ones to be destroyed. The result of this would be the formation of new species. Here then I had at last got a theory by which to work.

Darwin took Malthus's observation and applied it to his own observations of plant and animal species. He recognized that the reproductive potential of plants and animals vastly exceeds the rate necessary to maintain the population of these plants and animals at a constant level, yet such population sizes remain fairly constant. Based on this, Darwin concluded that the plants and animals that survive the intense competition among all life must be better equipped to live in their particular environment than the ones that do not survive. Predators battle for territory and prey, grazing animals seek the prime lands, plant life competes for space in which to expand and proliferate, males of many species fight for the right to the females. The swiftest, strongest, and smartest prevail and those vital characteristics are inherited by their descendants. Natural selection allows the changes that enhance survival to be passed on, and eliminates changes that do not. Though the science of genetics did not yet exist, Darwin surmised that essential qualities of the surviving predator, grazer, plant, and victor for the female were *somehow* passed on in their seed: "All these results . . . follow from the struggle for life. Owing to this struggle, variations . . . profitable to the individuals of a species . . . will tend to the preservation of such individuals and will generally be inherited by the offspring."

That was his proposition, *his theory of evolution!* It began as a "theory," but in the following decades, scientists would unequivocally prove the following facts as stated by Darwin: *The characteristics that contribute to each animal's superiority and success are preserved and passed on to future generations, each species gradually adapting and continuing to do so as the environment changes, in terms of climate, geography, available food, predators' adaptations, and all other aspects of that living thing's surroundings.* His answer to the question he initially posed on Galapagos was brilliant in its simplicity: plant and animal species undergo change because the most fit members of a species survive the gradually changing environmental elements. They adapt. They evolve along with their surroundings. If they fail to do so, they die, and if enough die, they eventually become extinct. Therefore, most extinct species are simply the earlier form of their progeny that flourish today. Though there are species that died off in massive numbers, like the dinosaurs, most fossils represent extinct species that did not actually abruptly disappear or die off without successors or offspring—they simply evolved over millennia into later and "better" models.

There are presently about 7 million species in the kingdom Animalia (including about a million species of insects)—millions of "better" models that have evolved to survive in their own environment and specific niches (places for that species to live successfully). So if individual members with favorable naturally selected traits are free to breed with all members of the same species, how did natural selection originally result in this great proliferation of species? The answer lies in "speciation"—the ability of a species to separate and subdivide into other species—and particularly one aspect of speciation called "reproductive isolation." Reproductive isolation can occur in a number of ways, usually because of geographical or physical developments that result in barriers or distance between members of the same species: rivers change course, mountains form, and continents drift apart (see Chapter 15). Also, groups of individuals in a species often strike out or migrate seasonally into new areas in search of a better food supply or environmental conditions. Any of these developments can result in a degree of isolation that will take individual organisms and species groups down a path that (combined with mutation and natural selection) brings about incompatible genetic makeup between the two formerly like models. That is, the two groups can no longer reproduce and create fertile offspring. Where one species existed, there are now two that are reproductively isolated. Where there were two, after millennia there could be four and then many more. Thus one common ancestor is shared by foxes, wolves, coyotes, and dogs, all prominent members of the family Canidae. Similarly, *Eohippus* (the genus of the "Dawn Horse") branched off over a 50-million-year period into rhinos, tapirs, zebras, donkeys, and modern horses.

The same principles of natural selection apply to all organisms, from the smallest one-celled bacteria to the largest whales. For example, certain types of mites, which are microscopic spider-like creatures, make their homes in the lungs of monkeys, noses of seals, and the breathing tubes of honey bees. Remarkably, mites of the species *Demodex folliculorum* live inside the pores

FIGURE 13-3

A. A Common Hard Tick (Family Ixodidae), a Type of Mite

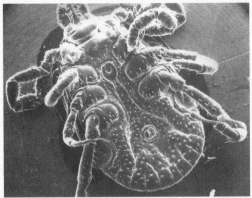

A. R. Brody

A hard tick (ticks are species of mites), *Dermacentor variabilis* (the dog tick), of the family Ixodidae, is adapted to feed on the blood of animals and people. See enlarged mouthparts in panel C. Most of the mites are free living (nonparasitic) like this example of a soil mite, *Oppia coloradensis* (panel 13-2B), which lives in the leaf and pine litter of forest floors. This species is found in the mountain forests of Colorado, but other similar species are found worldwide where they feed on fungal spores and a variety of organic materials. The eight legs of the dog tick and soil mite show the mites' relationship to spiders. The actual size of this individual soil mite is about 0.5 millimeter (500 microns). The tick is more than three times larger.

B. *Oppia Coloradensis*

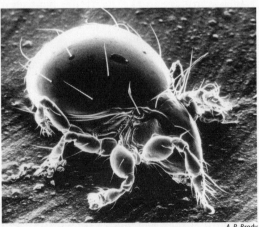

A. R. Brody

C. Tick Mouthparts

A. R. Brody

The hard ticks have mouthparts adapted to pierce the toughest skin and pump blood that is digested as their primary food source. If you look with a scanning electron microscope at the head end of the tick seen in panel A and view the tip of its suction tube, you see numerous serrated teeth that keep the tick securely tethered to its host while feeding.

of the skin covering the eyelids and faces of many of us. They eat, mate, and lay eggs in these minute, but obviously ideal, ecosystems.

There are thousands of different kinds of mites and each has adapted to a specific niche in the ecosystem in order to survive. Having mites in your skin is not a sign of poor hygiene—it is an example of an organism's successful adaptation to an available niche. These mites evolved hundreds of millions of years before humans existed, as evidenced by certain species of mites that have been preserved in amber dating back to the Paleozoic Era (545–251 million years ago). Before humans evolved, *Demodex* lived in skin pores of other animals such as dogs and sheep, and still live with those other species. The mouth parts of mites have adapted to pierce, suck, tear, and chew, depending on the niche in which the species is found. The tick shown in Figure 13-3A, viewed with a scanning electron microscope, has a tube with serrated edges that can pierce the toughest skin (Figure 13-3C). Upon entry into the skin, the tick secretes a fluid that prevents blood from coagulating, thus allowing the blood of its human or animal host to be pumped into the tick's stomach. When engorged with blood, the tick releases, digests its meal, and waits for the next passerby. It is fascinating to know that there are more than 40,000 different species of mites, and most of them are free living, roaming the forest floors, swimming in ponds, and feeding on your house plants. Others live inside the shafts of birds' feathers and the breathing tubes of honey bees. An example of a common mite that feeds on mold spores is seen with an electron microscope in Figure 13-3B. The evolutionary adaptation of mites is a story repeated for every species dispersed across this planet.

Each of those organisms has no "choice" but to develop the characteristics most necessary for survival. Some courses taken by certain species negatively impact other species, as we will see in Chapter 16, in the case of infectious bacteria. On the other hand, natural selection in the Fungi kingdom led to one of the greatest discoveries in medicine. The antibiotic penicillin is made from the colorful mold that commonly grows on bread. This mold is a species of fungus (*Penicillium*) that adapted to life by producing molecules that kill many bacteria. The production of the substance essential for penicillin is simply the fungi's well-adapted defense against bacterial infection. This and numerous other examples of natural selection have greatly benefited humankind.

Darwin imagined the branches of the rich and widespread tree of life pulling itself inward as the evolutionary clock runs in reverse over hundreds of millions of years. He speculated that all animals have "one common ancestor" and called his principle *natural selection* or *survival of the fittest.* "Natural Selection leads to the improvement of each creature in relation to its . . . conditions of life." Darwin correctly surmised that evolution is not random (except for randomly occurring genetic mutations, about which he could not have known; see Chapter 19) and doesn't depend on chance but does rely on the multiple powerful influences of natural selection that, in turn, are determined by geology, random climatic and competitive events, and other forces that make up the total environment and that have been shaping forms of life since the dawn of time.

After Twenty Years of Work, Darwin Publishes *On the Origin of Species*

An Ape for a Grandfather

On January 29, 1839, Darwin married his cousin, Emma Wedgwood, and began a family that was to eventually number ten children, several of whom were to be distinguished in their own careers, including three who were knighted. Darwin supported the family by selling books and articles on biological subjects related to his collections and observations during the H.M.S. *Beagle* voyage. During the years after the voyage, in a manner and outlook similar to the way Newton developed the basic laws of physics in the years after Woolsthorpe, Darwin secretly continued to compile his case for natural selection: "After five years' work I allowed myself to speculate on the subject, and drew up some short notes; these I enlarged in 1844 into a sketch of the conclusions." He shared his theory with a few close friends, including geologist Charles Lyell, who were skeptical at first but eventually were persuaded. Because of Darwin's humble and cautious nature and Lyell's advice "never to get entangled in a controversy," Darwin's original plan had been to leave the manuscript upon his death with instructions to his wife on getting it published:

> I have just finished my sketch of my species theory. . . . I therefore write this in case of my sudden death, as my most solemn and last request, which I am sure you will consider the same as if legally entered in my will, that you will devote four hundred pounds to its publication and . . . take trouble in promoting it. I wish that my sketch be given to some competent person, with this sum to induce him to take trouble in its improvement and enlargement. . . . Mr. Lyell would be the best if he would undertake it; especially with the aide of Hooker .

Darwin had feared the inevitable uproar that would be created by such a theory. Thus, in Newton-like procrastination, he would not have publicly disclosed this revolutionary concept had it not been for the writings and similar views of the British naturalist Alfred Russel Wallace (1823–1913), which caused Darwin to come forth and explain:

> My work is now (1859) nearly finished; but as it will take me many more years to complete it . . . I have been urged to publish this Abstract. I have more especially been induced to do this, as Mr. Wallace . . . has arrived at almost exactly the same general conclusions that I have on the origin of species.

Prompted by Wallace's competing work, Darwin finished his monumental book, *On the Origin of Species by Means of Natural Selection or the Preservation of Favoured Races in the Struggle for Life.* It was first published on November 24, 1859—more than twenty-two years after his first notes on the subject—and sold out immediately. It was republished in revised editions, with the sixth and last edition appearing in 1872. He gave credit to those who came before him and who set the stage for his own work. But it was Darwin's biological evidence that strongly suggested that existing species have evolved from earlier different forms, and it was Charles Darwin, not those he acknowledged, who figured out how natural selection produces adaptation, to the extent it could be logically explained without knowing about genetics, DNA, or the detailed workings of cells.

From the moment of its initial publication in 1859, the book became the subject of intense and widespread controversy. For the next several years Darwin maintained a low profile while his friends and colleagues—Joseph Hooker, T. H. Huxley, Alfred Russel Wallace, and Charles Lyell, a group of scholars and professors who were respected and distinguished in their own right—manned the front line in the war against mythical beliefs.

Calling him "the most dangerous man in England," Darwin's critics and enemies comprised a diverse group. English geologist Adam Sedgwick (1785–1873), a professor at Cambridge and president of the prestigious Geological Society of London, never gave a sound reason for his opposition. He would say only that he rejected Darwin's use of hypotheses and deductive (as opposed to inductive) reasoning to prove his theories. Richard Owen (1804–1892), a British anatomist, paleontologist, fellow of the Royal Society, and the nation's leading biologist at the time Darwin put forth his theory, was Darwin's most vehement and vicious detractor. A jealous and ambitious man, Owen unscrupulously attacked Darwin on numerous occasions. Many others who tried to discredit Darwin and his theory did so on the basis of their religious beliefs. If evolution were true, the account of the creation in the Book of Genesis was false. To their further dismay, if natural selection worked automatically, there was no divine guidance in the life and growth of plants and animals.

These issues were debated at a meeting of the British Association for the Advancement of Science on June 30, 1860, in Oxford, with Owen's spokesman, Bishop Samuel Wilberforce (1805–1873), on one side, and Darwin's spokesman, T. H. Huxley (1825–1895), on the other. This confrontation symbolized the firm division between science and theology. In response to Wilberforce's patronizing comment during the debate, concerning Huxley's relation to apes and monkeys, Huxley gave his famous response:

> If . . . the question is put to me, would I rather have a miserable ape for a grandfather or a man highly endowed by nature and possessed of great means of influence, and yet who employs these faculties and that influence for the mere purpose of introducing ridicule into a grave scientific discussion—I unhesitatingly affirm my preference for the ape.

Amid Growing Controversy Darwin Applies Evolution to Humans

Slits on the Side of the Neck

Contrary to popular belief, and even though Huxley and Wilberforce debated human evolution just seven months after publication of *On the Origin of Species*, that book does not address the evolution of humans. It's about plants and animals. In fact, there is just one brief comment by Darwin on *human* evolution, which appears toward the end of the book: "In the future I see open fields for far more important researches. . . . Much light will be thrown on the origin of man and his history." Darwin's view on whether his theory also applies to humans was written between the lines, but he left it to other biologists and the public to raise the issue: "During many years I collected notes on the origin or descent of man, without any intention of publishing on the subject, but rather with the determination not to publish, as I thought that I should thus only add to the prejudices against my views."

In the mid-1800s, the belief that we are immutable and unchanging was as clear as the universally accepted "fact" in the mid-1500s that the sun circles the Earth. To question the essence of our very being was an insult even greater than attacking the significance of our place in the universe. Yet that was the inevitable impact and result of *On the Origin of Species*. Until Darwin, humans were something special. Until Darwin, humans were qualitatively different from animals. This Earth and this universe were created by God for man and no other reason . . . until Charles Darwin.

Twelve years after publication of *On the Origin of Species*, controversy boiled anew. Darwin was compelled to pick up his pen once again to squarely address the subject of human evolution in *The Descent of Man*, published on February 24, 1871. Yet Darwin had been contemplating the issue at least since November 27, 1838, the date of his first notes applying natural selection to humans. In *The Descent of Man*, he explained how he'd abandoned his original intention not to become embroiled in the controversy: "Now the case wears a wholly different aspect," he wrote, citing the growing acceptance by respected naturalists of his basic thesis of natural selection. He felt the call "to put together my notes, so as to see how far the general conclusions arrived at in my former works were applicable to man."

The two-volume work totals nearly a thousand pages and is even more ambitious than *On the Origin of Species*. "The sole object of this work is to consider, firstly, whether man, like every other species, is descended from some pre-existing form; secondly, the manner of his development; and thirdly, the value of the differences between the so-called races of man."

Buoyed by the acceptance of his theory of evolution by natural selection, Darwin had grown bold in the twelve years that separated his two greatest books. In *The Descent of Man* he asserted that "man is descended from some less highly organized form. . . . The great principle of evolution stands up clear and firm, when these groups of the facts are considered . . . such as the nature of the affinities of the members of the same group, their geographical distribution in past and present times, and their geological succession." Regarding that geographical distribution, he predicted that the oldest fossil evidence of modern human's predecessors would someday be found in Africa:

> In each great region of the world, the living mammals are closely related to the evolved species of the same region. It is, therefore, probable that Africa was formerly inhabited by extinct apes closely allied to the gorilla and chimpanzee; and as these two species are now man's nearest allies, it is somewhat more probable that our early progenitors lived on the African continent than elsewhere.

Because of the lack of fossil evidence during Darwin's lifetime and because the "Dark Continent" was considered unfit as a place of man's origin, this prediction was considered folly all the way into the mid-twentieth century. Yet his fundamental theory was soon widely accepted:

> Man is developed from an ovule, about the 125th of an inch in diameter, which differs in no respect from the ovules of other animals. . . . The embryo itself at a very early period can hardly be distinguished from other members of the vertebrate kingdom. . . . [T]he slits on the side of the neck still remain . . . and the os coccyx projects like a true tail extending considerably beyond the rudimentary legs. . . . We thus learn that man is descended from a hairy quadruped, furnished with a tail and pointed ears, probably arboreal in its habits . . . probably derived from an ancient marsupial animal, and this through a long line of diversified forms, either from some reptile-like or some amphibian-like creature, and this again from some fish-like animal.
>
> Man may . . . feel . . . some pride at having risen . . . to the very summit of the organic scale. . . . But we are . . . concerned . . . only with the truth as far as our reason allows us to discover it. I have given the evidence to the best of my ability; and we must acknowledge . . . that man with all his noble qualities, with sympathy which feels for the most debased . . . with his god-like intellect which has penetrated into the movements and constitution of the solar system—with all these exalted powers—Man still bears in his bodily frame the indelible stamp of his lowly origin.

Fossil Records Confirm Evolution

A Fondness for Beetles

Darwin created the foundation that biologists, paleontologists, and geologists later built on to explain the evolution of all life. The science of geology established "eras" primarily to reflect major changes in the development of life on Earth by designating geologic time spans of great magnitude. The eras are divided into periods, which are further broken down into epochs. We are presently in the Holocene Epoch (which began 10,000 years ago) of the Quaternary Period (which goes back 2.5 million years) of the Cenozoic Era (which covers the last 65 million years). Here is a summary of the evolutionary history of all life as revealed by the fossil evidence, divided into each of the four geologic eras:

Precambrian Era (4.6 billion to 545 million years ago): Replicating life forms containing nucleic acids that code for proteins emerge from the primordial soup about 4 billion years ago. For the next 3 billion years, prokaryotes colonize Earth and remain the only life. About 600 million years ago, the first multicellular life evolves.

Paleozoic Era (545 to 251 million years ago): An incredible explosion of life begins 545 million years ago with the diversification of underwater life, leading to shelled animals 425 million years ago, followed by land animals 395 million years ago, the evolution of amphibians from fish about 350 million years ago, and reptiles beginning 280 million years ago.

Mesozoic Era (251 to 65 million years ago): After the most devastating extinction in history, 250 million years ago, the dinosaurs evolve and dominate the Earth from 230 to 65 million years ago, then die out.

Cenozoic Era (65 million years ago to present): After another mass extinction, the Earth's climate cools and vast rain forests give way to grasslands and woods. Mammals diversify into numerous forms. About 4 million years ago, the first members of the human family of primates climb out of the African trees.

Starting about 545 million years ago (during the Cambrian Period of the Paleozoic Era), evolution experienced its own "big bang"—there was an explosive radiation of life in which the predecessors of nearly all the species that now blanket Earth (and others that are now extinct) first appeared. The famous biologist Stephen Jay Gould (1941–2002) wrote about the Cambrian explosion in his book *Wonderful Life: The Burgess Shale and the Nature of History*, which explores the pace and breadth of evolution. Until recently, paleontologists and biologists believed that the Cambrian explosion of life lasted anywhere from 20 million to 70 million years. However, in 1993, a team of scientists from Harvard University, MIT, and the Yakutian Geoscience Institute in Yakutsk, Russia, found evidence in Cambrian rocks indicating that the period of rapid diversification lasted a much shorter time, perhaps only 5 million years or less. Gould was a proponent of what he called "punctuated equilibrium," which states that evolutionary change proceeds in "fits and starts," not smoothly through millennia as thought by many evolutionary biologists. There is no doubt that some species demonstrate punctuated equilibrium and rapid change (Darwin's finches, for example), whereas others remain largely unchanged for millions of years (crocodiles).

A great diversity of plant and animal life forms flourished for over 200 million years, and then suffered a massive setback about 250 million years ago when *90 percent* of all ocean life and a large percentage of land species were destroyed—the most massive and devastating extinction ever recorded and the beginning of the process that allowed for the ascension of the dinosaurs. The fossil record reveals that in some areas as much as 97 percent of all trees had died by the end of the Permian Period (250 million years ago), resulting in a proliferation of wood-rotting fungi that virtually covered the planet. Some paleobotanists and other scientists believe that a series of volcanic eruptions in Siberia caused this extinction. The sediments show that over a 1-million-year period, one volcano in particular poured out the equivalent of so much molten basalt that it could have paved the entire Earth with a layer six yards deep. The carbon dioxide and acids released from such eruptions could have poisoned the air and water and blocked the sun, thus causing the extinctions.

In 1996, four researchers proposed a different theory—that a huge upwelling of deep ocean water laden with carbon dioxide caused the Permian catastrophe. These biologists and geochemists maintain that this dangerously high level of carbon dioxide was natural at the ocean depths but entered the surface sea water and the atmosphere over a period of many years because of a "global turnover of deep ocean water," as now reflected in the pattern of calcium-carbonate sediments laid down at the end of the Permian Period. These excessive amounts of carbon dioxide would have caused carbonic acid to form in animals' blood in lethal doses and could have produced a greenhouse effect across Earth, destroying marine animals in massive numbers and upsetting many other species' delicately balanced environments.

The mass extinction at the end of the Permian Period gave rise to the dominance of the dinosaurs for about the next 140 million years. The dinosaurs were wiped out by yet another extinction (at the end of the Cretaceous Period), most likely caused by a massive asteroid or comet that landed near the Yucatan peninsula, resulting in a shroud of dust that encircled the globe and wreaked havoc on the climate and vegetation on which the dinosaurs relied, restricting their habitat and squeezing them to extinction. But this extinction opened up new ecological niches. With the forces of natural selection at work, life rebounded with greater success and diversity.

The latest burst of life, following extinction of the dinosaurs, surrounds us. It is seen by examining a pinch of soil or by looking at the vast number of species that now inhabit Earth. With the aid of a microscope you would find about 10 billion of the longest and most enduring residents of planet Earth: *bacteria*. These one-celled organisms (representing hundreds of species in those 10 billion bacteria) break down the organic material in the soil. Overall, biologists have identified and named close to 2 million species of animals, plants, and microorganisms, although most experts estimate that literally millions of species of bacteria and insects, and certainly some fish, birds, and mammals, have not yet been discovered. As explained in Chapter 12, biologists expanded Carl Linnaeus's original two-kingdom scheme (plants and animals) into today's six-kingdom system: (1) Archaea and (2) Eubacteria (true bacteria) (both primative microorganisms without a complete nucleus), and the Eukaryotes with a true membrane-bound nucleus; (3) Fungi (Latin for "mushrooms"); (4) Protists or Protozoa (meaning "first animals"); (5) Plantae; and (6) Animalia.

The following list shows the large number of insect species in comparison with all other life forms, including the extraordinary number of beetle species. When asked what the study of creation had revealed about God's plan, Darwin's friend T. H. Huxley quipped, "That the Almighty has an inordinate fondness for beetles."

Known Species

- 5,000 bacteria species (1,000 species on and in the human body alone)

- 31,000 protozoan species

- 80,000 fungi species

- 320,000 plant species

- 1,000,000 insect species (of which 290,000 are beetles)

- 305,000 species of invertebrates (without a backbone) like corals, shellfish, mites, and spiders

- 60,000 species of vertebrates like fish, amphibians, reptiles, birds, and mammals (including our single human species)

At the end of *On the Origin of Species*, Darwin wrote, "There is a grandeur in this view of life. . . . Whilst this planet has gone cycling on according to the fixed law of gravity, from so simple a beginning endless forms most beautiful and most wonderful have been and are being evolved."

Indeed, a number of scientists have followed Darwin to the Galapagos Islands to further understand how the unique niches there shape species' characteristics. In 2008, the scientist couple Rosemary and Peter Grant culminated thirty years of studying Darwin's finches with a remarkable discovery on how quickly evolution can proceed by natural selection, the so-called engine of evolution. Even though Charles Darwin had not even labeled the various birds he had taken from the islands, about fourteen different species (more or less) carrying his name populate the Galapagos Islands today. In their article "How and Why Species Multiply," the Grants describe their extensive, tedious leg-banding and nest-counting techniques and measurement of the beaks of numerous birds. Large robust beaks (like that of the large ground finch shown in Figure 13-3A) crush seeds easily, but these birds are at a disadvantage when the seeds available are small. The Galapagos Islands regularly see large climate shifts controlled in large part by El Niño events in the Pacific Ocean, and vegetation is altered accordingly, with prolonged drought leading to large tough seeds and heavy rain producing dominant plants with small seeds. Could the Grants demonstrate the evolution of those Darwin's finches that were successful at feeding on the seed types available? Would changes in microclimates and the consequent seeds that developed select for beaks that would be expected if evolution were proceeding by natural selection? This is precisely what they found. Each climate change over thirty years resulted in shifts in beak shapes and body sizes. Figure 13-3B shows the thin narrow beak of a tree finch that has adapted to feed on the small seeds. The Grants demonstrated colonization of individual species and the development of individual barriers to population interbreeding on several of the islands. This of course is much faster than most of the evolutionary changes we are used to seeing, but should anyone doubt that evolution by natural selection is an established concept in biology and is operating as we live and breathe, they should read not only the work by the Grants but also Jerry Coyne's book *Why Evolution Is True*, in which he explains in beautiful detail how evolution appears as both a theory and a fact. He knows that intense disagreements about what is theory and what is fact is not a scientific debate, but rather is social controversy among those who understand evolution and those who choose not to.

In *The Origin*, Irving Stone wrote that at the end of Darwin's life:

> Perhaps it was the long bouts of illness; perhaps the concentrated periods of incredibly hard work; perhaps it was his snow-white, all-encompassing beard . . . but Charles Darwin now looked as though he were the Patriarch of the World. His eyes were sunken; he carried the look of a man who was about to say farewell; and did not mind at all.

Throughout his life Darwin had suffered from a number of ailments, including episodes of depression, and a heart attack in 1873. On April 9, 1882, he suffered another heart attack and died. Upon petition by twenty members of Parliament, Charles Robert Darwin was laid to rest in Westminster Abbey.

He had come a long way from the poor student whose father feared he would be "a disgrace to the family," and from the collector of specimens on the *Beagle*. He evolved from the Great Describer to the Great Interpreter. He kept asking until there was an explanation that made sense. He collected the unordered facts available and made them coherent and understandable. In his own lifetime he not only authored evolutionary theory but succeeded in changing the public's opinion of it from an outrageous heresy to a generally accepted principle. Entire scientific disciplines have grown directly out of Darwin's work. The field of genetics (discussed in Part Six), and the search for the ultimate mechanism of heredity and evolution—the DNA molecule (Part Seven)—have now combined with Darwin's work to form a complete picture of evolution, resulting in a fundamental change in our concept of all life, including the human race.

Chapter Fourteen
SHAPER OF THE LANDSCAPE

Every animal leaves traces of what it was. Man alone leaves traces of what he created. . . . Man is not a figure in the landscape. He is a shaper of the landscape.

How did the hominids come to be . . . dexterous, observant, thoughtful, passionate, able to manipulate in the mind the symbols of language and mathematics both, the visions of art and geometry and poetry and science? How did the ascent of man take him from those animal beginnings to that rising enquiry into the workings of nature, that rage for knowledge?
—Jacob Bronowski, *The Ascent of Man* (1973)

Charles Darwin formulated the theory of evolution by natural selection. But in his day no one had any idea how long it had taken for humans to evolve or through what stages it had occurred. Darwin opened the book on the evolution of humans. Others continue to write the story. The chapters that anthropologists and paleoanthropologists have written in this century represent yet another turning point in the history of human knowledge and in our concept of our primitive ancestors.

Diverse Fields of Science Combine in the Twentieth Century to Define the Course of Human Evolution

Mammals Thrive on December 30

Paleoanthropology is the study of early humans, including fossils of hominids (human-like beings) and extinct primates from which we evolved, as well as relics of the cultural activities of ancient humans. As we learned in Chapter 13, dinosaurs became extinct about 65 million years ago when the Earth's climate cooled and mammals began to flourish across the Earth. With the wealth of physical information discovered in recent decades, together with the application of unprecedented technology, twentieth and twenty-first century paleoanthropologists have connected the human and nonhuman links of this 65-million-year-old chain by combining developments in several disciplines.

- *Physics and geology:* Through twentieth-century developments in physics (radioactive dating) and geology (geological stratification), we are now able to measure the age of fossils with great accuracy.

- *Biochemistry and biology:* In Parts Six and Seven we will also see how biochemists and biologists discovered how the fundamental "code" of reproduction was passed from primitive life to humans.

- *Astrophysics and geochemistry:* Even astrophysics and geochemistry have helped put pieces of the evolutionary puzzle together through the study of the formation of elements in space and the study of the chemical composition of the early Earth.

Thus we've been able to re-create the evolutionary process of humans through a remarkable synthesis of diverse disciplines that were once thought to be unrelated. As Darwin predicted, the process began with "one common ancestor" well before the appearance of reptiles and amphibians. This evolutionary path has been ongoing for billions of years. To help conceptualize the history of evolution, we will squeeze the 14-billion-year history of the universe into an imaginary year, with the Big Bang occurring at the first instant on the morning of January 1 and the year 2000 being the first second after December 31. These would be the significant dates of that "Year of the Universe" *before* the month of December:

FIGURE 14-1

Year of the Universe—Through November

January through November (12.6 billion years)

January 1	Big Bang
May 1	Origin of Milky Way Galaxy
September 9	Origin of solar system
September 14	Formation of Earth
September 25	First life on Earth
October 9	Date of oldest fossils (bacteria and algae)
November 12	Oldest fossil of photosynthetic plant
November 15	First cells with nuclei

Since each month of that year represents about 1.14 billion years, the first life doesn't appear until September 25, and *recorded history doesn't begin until the last minute before midnight on December 31.* Here are the other significant events that occur during the 1.14 billion years of the month of December.

Sixty-five million years ago on this calendar is December 30. This is very recent, compared to the age of the Earth at 4.6 billion years and the 14-billion-year-old universe. Thus just "yesterday" the dinosaurs disappeared and human predecessors began to emerge. The hominid's tenure on Earth—a few million years—is substantially briefer than the 140-million-year reign of the dinosaurs. *In our modern form we've been around only about 40,000 years.*

Our Ancestors Are Revealed in the Fossil Records

Rodents in the Jungle of Europe

Sixty-five million years ago the world was covered with tropical forests from the equator to latitudes as high as what is now northern France and Germany and southern Alaska. A wide variety of plant and animal life thrived, including a rapid proliferation and evolution of mammals. Among

FIGURE 14-2

December in Year of the Universe

December

Sunday	Monday	Tuesday	Wednesday	Thursday	Friday	Saturday
	1 Oxygen atmosphere begins to develop on Earth	2	3	4	5	6
7	8	9	10	11	12	13
14	15	16 First worms.	17 (Paleozoic Era and Cambrian Period begin.) Invertebrates flourish.	18 First oceanic plankton. Trilobites flourish.	19 (Ordovician Period) First fish and first vertebrates.	20 (Silurian Period) First vascular plants. Plants begin colonization of land.
21 (Devonian Period begins.) First insects. Animals begin colonizaion of land.	22 First amphibians and first winged insects.	23 (Carboniferous Period) First trees and first reptiles.	24 (Permian Period begins.) First dinosaurs.	25 (Paleozoic Era ends and Mesozoic Era begins.)	26 (Triassic Period) First mammals.	27 (Jurassic Period) First birds.
28 (Cretaceous Period) First flowers. Dinosaurs become extinct.	29 (Mesozoic Era ends/Cenozoic Era and Tertiary Period begin.) First cetaceans and first primates.	30 Early evolution of frontal lobes in the brains of primates. First hominids. Giant mammals flourish.	31 (End of the Pliocene period. Quatenary Period begins.) First humans.			

them were the predecessors of the primates, including what are now called ancestral primates. They do not resemble their successors—the monkeys, the apes, the human beings—but over a period of 65 million years, it takes very small modifications by natural selection to turn a long-tailed rodent-like mammal (yes, your ancestor) into a totally different form of life. The closest relatives to our most ancient mammalian ancestor that still exist today are called prosimians (meaning early primate). Prosimians include lemurs, lorises, and tarsiers, and fossils have shown that these prosimians have changed very little over time, while other ancestral primates branched off and obviously went through radical changes.

Tracing human evolution from the time of that little, scuttling mammal is like watching an incredible slow-motion film. We can examine each stage of that 65-million-year-long story:

- Over the first few million years of the Paleocene Epoch (65 million to 54 million years ago), it evolves into a tree dweller and eats fruit and insects.

- Individuals with longer hind legs are naturally selected as the species adapts to a climbing, swinging, arboreal life.

- Flat nails replace rat-like claws, and it begins to develop a thumb that can be opposed to the hand.

- The fingers grow longer, become more flexible, and are able to grip.

- During the Eocene Epoch (54 million to 38 million years ago), the posture becomes more upright, the head can now rotate to a greater extent, and the brain grows.

- Eyesight improves and eyes grow larger and widely spaced, while the sense of smell decreases, indicating natural selection against smell and in favor of vision; accordingly, the snout becomes shorter.

- During the Oligocene Epoch (38 million to 26 million years ago), a branch of the prosimians begins to look much like monkeys or apes. They live in the tropical bush country of what is now Egypt.

- The teeth evolve into a four-cusp pattern unique to primates.

FIGURE 14-3

Paleocene and Eocene Ancestral Primates

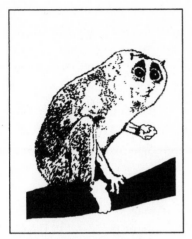

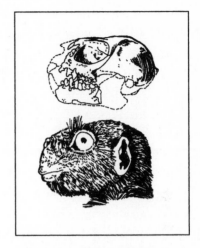

Loris found in Africa and Asia—Similar to Paleocene ancestral primate of 65 million years ago

Skull drawn from fossils of Eocene ancestral primate (top) of about 45 million years ago, and restoration of the head (bottom)

- The frontal bone above the eyes and around the eye sockets approaches closure at the back of the orbit, similar to that of humans.

- During the early part of the Miocene Epoch (26 million to 7 million years ago), the creature grows to about a foot tall, and continues to increase in size as it adapts to its environment.

- It evolves out of the trees and returns to the ground because tree fruit is less abundant. The diet changes to berries, roots, and insects. The animal begins to walk on the soles of its feet and knuckles of its hands but still has anatomical features that allow it to spend time in the trees.

- Skeletal adaptations occur consistent with the new mode of moving, eating, and living.

- The jaw becomes smaller and the face more vertical.

- These ancient hominids (with human characteristics), older than about 7 million years and existing through the Miocene Epoch, are in two main genera—*Sahelanthropus* and *Orrorin*. Numerous fossils from two species demonstrate that they look decidedly like small apes, but they have unmistakable human-like features suggesting that they walk more upright and are three or four feet tall. *Sahelanthropus tchadensis* and *Orrorin tugenensis* (the two species named so far) are hardly human, but they are clearly on this evolutionary track, perhaps as ancestors to both apes and humans as proposed by some paleontologists.

During the Pliocene Epoch (7 million to 2.5 million years ago), the developing hominid diverges (perhaps through *Orrorin*) from African apes like gorillas and chimpanzees and evolves into "ape man," as it develops a larger brain, retains an opposable thumb, and several other hominid characteristics, particularly the ability to walk upright on two legs. In 2009, identification of an almost complete skeleton of *Ardipithecus ramidus* from Afar, Ethiopia, was named the scientific "Breakthrough of the Year" in the December 2009 issue of *Science* magazine. "Ardi" as she was nicknamed, is hairy, and her remarkable opposable big toe shows she still is perfectly adapted to spend considerable time in the trees. *Ardipithecus* has flat feet (no arch) so she wouldn't have been too fast out in the open, but it appears from the portions of skeletons from about thirty-six different individuals that these hominids move in bands and are agile, keen-sighted, alert, and more intelligent than the baboons and apes with which they compete for food in the African savanna and forests. It does not have a language, but probably communicates with an array of sounds, gestures, and nonverbal expressions like all apes. Its life span is probably 15 to 20 years. More than 125 pieces of Ardi's skeleton have been identified and dated to about 4.4 million years ago. Her skeleton reveals the foundation for upright walking. Even though her body and brain are only marginally larger than a chimpanzee's, she does not look like either a modern or ancient ape (Figure 14-4A). Her face is flatter than the apes' and she lacks the huge upper canine teeth ubiquitous among apes. The base of her skull where the spinal cord emerges from the brain and the construction of her pelvis have convinced most paleontologists that Ardi walked upright, but her opposable big toe, more flexible wrist and ankle joints, and long curved fingers support the view that she could move comfortably among the tree branches as well (see these anatomical features in Figure 14-4A). If walking upright is the cardinal feature for being a hominid,

Ardi appears to be less well adapted for doing so compared to our more recent ancestors (introduced in the following paragraphs), and whether Ardi gave direct rise to *Australopithecus*, the closest genus to our own *Homo*, or whether she is an interesting side branch on the riveting path to *Homo sapiens* is under intense study. It is interesting to note that Ardi's existence was announced 200 years after Darwin's birth. Starting with Darwin, it took only 200 years for scientists to discover what appears to be a direct line from our ancestors 4 million years ago to where we stand today.

FIGURE 14-4

A. *Ardipithecus ramidus*

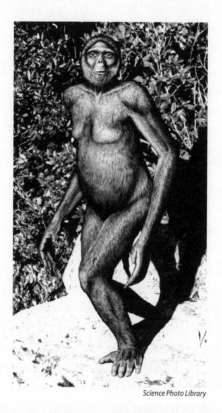

Science Photo Library

About 4.5 million years ago *Ardipithecus ramidus* traveled the forests and savannas of Africa. Numerous skeletal remains show that this early hominid walked upright, had an opposable big toe (seen in this image on the left foot), and long gracile fingers like apes for grasping branches but had lost the large canine teeth common to apes.

B. Australopithecine Attacked by Saber-Toothed Cat

About 2 million years ago, several *Australopithecus* species roamed the forests of southern Africa. Recent findings suggest that *Au. sediba* is a direct ancestor of our genus *Homo*. The australopithecine seen here is being attacked by a saber-toothed cat (see Figure 14-6).

The Pliocene Epoch (beginning about 5 million years ago) saw the arrival of *Ardipithecus*. Not quite halfway through the Pliocene, the best-described genus that walked the savannas and forests of Africa from about 4 to 1.5 million years ago has been named *Australopithecus*. The first of these to be discovered was *Australopithecus africanus* (The "Southern Ape of Africa" or Taung child) and the second, the icon of human evolution, *Australopithecus afarensis* ("Lucy"). Both of these will be discussed in further detail later. Paleontologists have found additional australopithecine species (for example, *Au. anamensis* and *robustus*), and through the following millions of years, it is clear that the australopithecines become less ape-like and more human-like. But which of our ancestors could fill the evolutionary gap between Lucy and the first of our own genus to be described, *Homo habilis*? In the April 9, 2010, and September 9, 2011, issues of *Science* magazine, a spectacular find was announced regarding *Australopithecus*. Careful dating procedures strongly suggest that *Australopithecus sediba* is a link on the path to the genus *Homo*. *Australopithecus sediba* roamed the savannas and forests of southern Africa about 2 million years ago, just about the time when our genus, *Homo*, was emerging. Scientists led by Lee Berger (1965–) from the University of the Witwatersrand in South Africa, found in caves in Malapa, South Africa, four separate individuals and described in great detail two separate skeletons showing that this species shares several traits found in *Homo* but not in Lucy, such as a precision grip provided by a human-like hand and several cranial features more like those of *Homo habilis*. Some paleontologists would rather keep *Au. sediba* on the evolutionary path between *Au. africanus* and *Au. afarensis*, whereas Berger sees his remarkable discovery more as a transition from a hominid better adapted to the trees (*Au. africanus*) to a "full-striding" and upright *Homo erectus*. It is clear that throughout the Pliocene Epoch there were multiple species of primitive hominids ambling around southern and eastern Africa. Nature apparently was experimenting, by means of natural selection, with a number of ways to become a member of the genus *Homo*.

The chronology just described is primarily based on fossils discovered over the last seventy-five years. While there were earlier finds of ancient hominids, the first to be discovered and recognized as one of our ancestors was the Taung baby skull (named after the Taung limestone quarry), found by anatomist Raymond A. Dart (1893–1988) in 1924 in the Kalahari Desert, South Africa. It is more than 2 million years old and was the first concrete evidence found supporting Darwin's theory and fulfilling his prediction that fossils of ape-man would be discovered in Africa. Darwin said that if humans and apes are descended from the same ancestor, then evidence of human origins should be found in Africa because this is where gorillas and chimps, our closest living relatives, are found today. The Taung baby was identified as *Australopithecus africanus*. Similar fossils were found in 1936 and thereafter in the stalagmitic deposits of ancient caves and fissures formed in limestone in the Transvaal, South Africa. The original, beautifully fossilized Taung skull was passed on by Raymond Dart to the safe hands of his student Phillip Tobias (1926–2012) at University of the Witwatersrand. In Figure 14-5, the tiny child's skull is gently cradled by Tobias as several other hominid fossils rest on his desk. Tobias played a major role in identifying several species key to understanding human evolution, the most notable being the oldest member of our genus, *Homo*. There is more to learn about our genus, but now is a good time to consider Lucy (*Australopithecus afarensis*), thought by many to be on a direct line of evolution to *Homo*, at least until the recent discovery of *Australopithecus sediba*. It is interesting to know that all of the australopithecine fossils studied so far show a mixture of apelike and humanlike features. While it is clear the characteristics of their lower limbs leave no doubt that Lucy and her closest relatives walked upright, it is less clear as to what degree they climbed trees. In a 2012 article in *Science* magazine, an example of *Au. afarensis* from Ethiopia demonstrates a shoulder joint that would be most advantageous for tree climbing, much like the joints found in apes. Lucy has

FIGURE 14-5

Phillip Tobias Holding the Taung Baby Skull

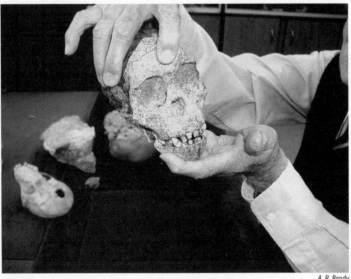

A. R. Brody

Renowned paleontologist Phillip Tobias cradles the priceless fossilized skull of the Taung baby (*Au. africanus*) discovered by Raymond Dart in 1924. It can be seen that *Australopithecus*, like *Ardipithecus*, has lost the large canine teeth common to all the apes.

this very same form of primitive shoulder joint, and Figure 14-4B is an artist's conception of an australopithecine ancestor trying to escape up a tree from attack by a saber-toothed cat. An actual skull of a saber-toothed cat that lived tens of thousands of years ago in South Africa is held by the author's (A. R. Brody's) wife (Figure 14-6) during a visit to Tobias's laboratory in 2005. The huge canine teeth that made this predator so effective for millions of years are obvious in the photograph. Very similar fossils of saber-toothed cats also have been unearthed in North and South America.

FIGURE 14-6
Saber-Toothed Cat Skull

A. R. Brody

Saber-toothed cats were dominant predators in Africa as long as 40 million years ago. The earliest hominids evolved about 8 million years ago, and the cats didn't become extinct until about 10,000 years ago, providing millions of years for hominids to live on earth simultaneously with the cats and to be taken as prey (see Figure 14-4B). This striking example of a skull from a saber-toothed cat that once lived in South Africa shows the massive canine (saber) teeth that made this predator so effective. The skull is held by A. R. Brody's wife, Toby, standing in Phillip Tobias's laboratory at the University of the Witwatersrand.

In 1959, anthropologist Mary Leakey found a 1.7-million-year-old complete *Australopithecus* skull in Olduvai Gorge in Tanzania, the first time such a skull was found outside of southern Africa. Ten years later, her son, Richard Leakey, found a similar cranium in a dry riverbed near the eastern shore of Lake Turkana in northern Kenya. In 1974, paleoanthropologist Donald C. Johanson (1943–) made the extraordinary discovery of an almost complete skeleton of a young girl (whom Johanson called "Lucy" because the Beatles song "Lucy in the Sky with Diamonds" was playing in the campsite) who lived 3.5 million years ago in what is now Ethiopia. Lucy became the basis for establishing the new species of human ancestor *Australopithecus afarensis*. This remarkable find resided in a museum in Ethiopia for decades but in 2007 took a trip to the Houston Museum of Natural Science so more of us could see her beautifully preserved skeleton. A painted cast of Lucy is on display at the Cleveland Museum of Natural History, where Johanson was working as a curator of anthropology at the time of his discovery. She continues her travels through the United States and in 2009 was on display in

FIGURE 14-7
Homo habilis

Science Photo Library

Homo habilis is the oldest member of our genus to be described. *H. habilis* clearly walked upright and used crude tools. *H. habilis* is the likely ancestor of the more human *H. erectus* (see Figure 14-8).

New York City. Many have thought that Lucy was the find of the century, but others are naming *Ar. ramidus* (see earlier discussion) as such because it is the oldest fossil on the direct line to our own genus, *Homo*.

Australopithecines are believed by many to have made crude stone tools and to construct shelters, indicating further biological changes in the hand and the brain centers that control the hand. About 100,000 individuals existed at any given time during the period when *Homo habilis*, the oldest described species of our genus, was thriving, from about 2.5 to 1.5 million years ago. In 1959, Tobias, then a young professor at Witwatersrand, received what he called a "cryptic telegram" from Mary and Louis Leakey, who had just unearthed what they believed to be a previously unknown hominid species at the Olduvai Gorge in Tanzania. The telegram said "We've found the man. . . . Come quickly." And indeed, there was *Homo habilis* (Figure 14-7), or the "Handy Man" as he was called because of his skills at honing and using stone tools. *H. habilis*, standing not much more than three feet tall, had smaller teeth and a larger brain than the australopithecines but still spent time in the trees and likely was common prey for a number of larger predators, like the saber-toothed cats. Tobias's painstaking description of the original *H. habilis* (calling him "Dear Boy') serves many paleontologists today as the model for how carefully a fossil should be described and depicted.

Homo habilis (Figure 14-7) is considered by many to be directly on the evolutionary pathway to the more human-like *H. ergaster* and *H. erectus*. Other paleontologists would like to see *H. habilis* as a more modern member of the genus *Australopithecus*. It is not known if *H. habilis* still had an opposable big toe, but based on fossilized footprints made by *Homo ergaster/erectus*, about 1.5 million years ago, our closest ancestors walked fully upright, on arched feet, and thus had finally departed from the treetops.

Over the course of the next several hundred thousand years, hominids begin to stand tall (about five feet). Their brain is now twice the size of that of *Australopithecus*. During the period beginning 1.5 million years ago and extending to 400,000 years ago, this new species, named *Homo erectus* (Figure 14-8), spans the transition between *H. habilis* and early *H. sapiens*. *H. erectus* is the first fully human species and the first hominid to leave Africa (about 1.8 million years ago) and begin to populate the rest of the Earth. It disperses across several continents and grows to a population of about 1 million at any given time.

In the twentieth century, *Homo erectus* fossils were found in China, Java, Africa, and Europe, all showing a cranial capacity of 850 to 1,000 cubic centimeters (sixty cubic inches), close to our own size range of 1,000 to 1,800 cc. They had a flat, retreating forehead, large browridges, flattened skull vault, large teeth and jaw, and small chin. They built huts, and the more recent ones (beginning 700,000 years ago) built fires and cooked their food. They invented tools for chopping wood, cutting meat, skinning animals, and scraping hides. There is new evidence reported in 2012 that by about 500,000 years ago, primitive man had developed a process called "hafting" where a stone point was attached to a wooden shaft, thus producing a projectile that we would recognize as a spear. Attaching such stone points to make more effective weapons was carried out by Neanderthals and early *Homo sapiens* as long as 300,000 years ago by inserting the point into a slit in the end of the shaft and binding the joint with leather strips wrapped around the end of the shaft. But the new evidence of such weapons hundreds of thousands of years earlier suggests that a common ancestor in Central South Africa,

FIGURE 14-8
Homo erectus

Science Photo Library

Homo erectus was the first hominid to leave Africa, and numerous fossils have been found across Africa and throughout the continents of the Old World. They clearly used tools and formed communities.

perhaps *Homo erectus*, had developed this effective technology that has persisted through modern societies.

At approximately the same time the species of *Homo* were extant in Africa another genus was thriving. This was *Paranthropus*, and at least three separate species have been identified. But *Paranthropus* was more robust and ape-like and apparently was a side branch of hominid evolution disappearing about halfway into the Pleistocene Epoch, 1 million years ago.

Homo erectus may have been able to speak, and develop a more complicated social organization than its ancestors. Labor was divided between the sexes, the family unit took on significance, bands of twenty to fifty individuals lived together, and permanent home sites were sometimes established. *H. erectus* could expect to live twenty to thirty years, not much less than its future relatives of fourteenth-century Europe, whose average age at death was thirty-eight.

The fossil record of *Homo sapiens*'s ancestors is becoming clearer. Of course there are gaps and many questions that paleontologists and anthropologists are working hard to fill. This is a fascinating and rewarding field of discovery that requires dedication and hard work. And because of the successes of many, we are learning that our early ancestors who wandered the plains on two legs were physically suited for a nomadic life. Those who created the most successful hunting strategies and found the safest havens for their family were the most likely to survive. The strongest, fastest and, foremost, the most intelligent members of the tribe or band were more likely to pass their genes on. Just as the fastest and fiercest lions prevail among that species, the most intelligent hominids and humans prevailed among our ancestors and developed into modern *Homo sapiens*.

Returning to the Year of the Universe model again, here are the events that immediately preceded the present. The year 2000 would begin the first second after December 31, so the following time table identifies the significant events of the "afternoon" and "evening" of December 31:

FIGURE 14-9

December 31 of Year of the Universe

Origin of prehistoric ancestors of humans	1:30 p.m.
First humans	10:30 p.m.
Widespread use of stone tools	11:00 p.m.
Beginning of last glacial period	11:56 p.m.
Invention of agriculture	11:59:20 p.m.
First cities	11:59:35 p.m.
First dynasties in Egypt	11:59:35 p.m.
Development of astronomy	11:59:35 p.m.
First writing	11:59:51 p.m.
Birth of Christ	11:59:56 p.m.
Fall of Rome	11:59:56 p.m.
Renaissance in Europe	11:59:59 p.m.
The twenty-first century	The first second of New Year's Day of 2000

We'll now examine paleoanthropology's finds that have given us a clear picture of the humans (genus *Homo*) who evolved beginning at about 10:30 p.m. on this last day of the Year of the Universe.

Neanderthal Man Emerges, Then Disappears

Rickets, Arthritis, and the Newcomers

In 1856, in a cave near Dusseldorf, Germany, a science teacher by the name of Johann Carl Fuhlrott found fourteen pieces of bone that looked like those of a distorted human. The skull was long, low, and wide. It suggested that this being's face had heavy browridges, a receding forehead, small cheekbones, and large teeth. The chest was broad, and the limbs were heavy, with curving thigh and forearm bones, larger-than-human feet and hands, and short fingers and toes.

This was the first ancient human fossil skull ever to be discovered. It was named Neanderthal, after the Neander Valley, Germany, where the bones were found. The first edition of *On the Origin of Species* appeared three years later, in 1859. But even then, no one was prepared to accept what appeared to be a primitive animal as his or her ancestor or to accept the idea that early humans lived in caves. Fuhlrott himself concluded that the bones were probably distorted from the weather and that they belonged to some unfortunate person who was washed into the cave by Noah's flood. Other scientists who examined the skull specimen concluded that it was a very ancient pathological idiot. The highly respected German anatomist, pathologist, and statesman Rudolph Virchow (1821–1902), wrote that the bones were from a Mongolian cavalryman who suffered from rickets in childhood and arthritis in old age, and who had been clubbed on the head. Even Darwin's friend and supporter Thomas H. Huxley, the world-renowned

FIGURE 14-10

Homo neanderthalensis

Science Photo Library

Homo neanderthalensis was the dominant hominid for hundreds of thousands of years and became extinct about 30,000 years ago.

biologist, studied the find and said, "In no sense can the Neanderthal bones be regarded as the remains of a human being intermediate between men and apes." This effectively silenced all speculation for the time being.

In 1886 two more Neanderthal skeletons were found, this time near Spy, Belgium. Primitive stone tools and remains of extinct animals were discovered alongside the skeletons. Though Virchow repeated his theory about diseased specimens, many scientists were beginning to recognize that the idea of primitive ancestors was valid, although distasteful. More complete Neanderthal fossils were found in France, beginning in 1908. In subsequent years, remains were found throughout Europe, Asia, and Africa, and now total about 300 individuals.

Today, most paleoanthropologists agree that our ancestors and Neanderthals diverged somewhere between 250,000 and 450,000 years ago. *Homo sapiens* evolved along a line from *Homo erectus*, the first hominid to migrate from Africa about 1.8 million years ago. But today it remains unclear just which archaic *Homo* is the direct descendant of *Homo neanderthalensis*. Relatively modern *Homo* species such as *H. rhodesiensis* and *H. heidelbergensis* appeared in the Holocene Epoch (beginning about 1 million years ago) contemporary with Neanderthals, and display several anatomic characteristics of Neanderthals. These powerful, resourceful hominids were the dominant species of *Homo* for hundreds of thousands of years on the Eurasian continent until they were superseded by *Homo sapiens* and vanished about 30,000 years ago. Neanderthal bones have played a key role in studying the evolution of humans. These people may have been nearly as intelligent as modern humans, but this is unlikely given how many millennia they existed with a concomitant lack of technical and cultural progress. They had a three-pound brain, which is as large as the brain of modern humans and stood as erect as *Homo sapiens* (Figure 14-9). Surviving in the harsh and frigid wilderness of intermittent Ice Ages, they were cave dwellers, used fire, including torches and lamps, hunted with spears, and occasionally caught fish, using a variety of tools made from stone, wood, and bone. They also buried their dead, spoke a rudimentary language, and might have practiced a primitive religion. Fossils contain significant evidence of cannibalism. They lived in groups of up to thirty individuals and roamed as far north as Britain and as far south as Spain, as well as areas in Asia and Africa where multiple remains have been found. The average age at death was thirty, and their total population was in the tens of thousands at the height of their existence.

These early modern humans were contemporaries of *Homo erectus* in Eurasia. They were the individuals generally regarded to be the Stone Age humans. Neanderthals suddenly disappeared from the fossil record and were replaced by a new, more advanced culture that either annihilated them by force or out-competed them for available resources by means of superior intelligence. The "newcomers" were different from the Neanderthals. They had a high forehead, a more pronounced chin, and a flatter face. Anatomically, they were the same as modern *Homo sapiens* and would be known as "Cro-Magnon." *But the original* Homo neanderthalensis *is not entirely lost to time!* As reported in the May 7, 2010, issue of *Science* magazine, paleontologists studying Neanderthal bones were able to extract sufficient DNA to announce that Neanderthals interbred with our ancestors who developed further into the varied population groups of Europe and Asia. Modern humans share as much as 4 percent of their nuclear DNA (see Chapter 19) with *H. neanderthalensis*, as long as that human is not directly evolved from an African ancestor. This means that Neanderthals interbred with modern humans after *H. sapiens* migrated from Africa. This could be as early in the Holocene Epoch as 80,000 years ago, but it is clear that Neanderthals coexisted with modern humans (in Eurasia not Africa) from about 30,000 to 45,000 years ago. Our genome is 99.84 percent the same as the Neanderthals' (see discussion of the Human Genome Project on page 332), and scientists have identified at least eight different genes (among many to be sure) that make us *H. sapiens* and not *H. neanderthalensis*.

In 2011, another modern human group appeared on the scene, the Denisovans. The renowned paleogeneticist Svante Pääbo, from the Max Planck Institute in Leipzig, Germany, recovered sufficient DNA from a remarkable place in southern Siberia called Denisova Cave to establish that previously undescribed archaic humans, whom he called the Denisovans, lived from about 30,000 to 50,000 years ago. The species has not yet been named, but Pääbo states that Denisova Cave is the one place on earth that scientists are sure that the three human forms, Neanderthals, Denisovans, and *H. sapiens*, all lived (probably not at the same time) at some relatively recent time in history. There is cultural evidence for this conclusion in the form of sophisticated stone tools as well as Denisovan teeth that could not have come from a modern human or a Neanderthal. It looks to Pääbo like the Denisovans also had a wide range of habitats because their DNA has recently been reported in well-characterized populations like Australian Aborigines and Melanesians, and these individuals have about 5 percent of their DNA from the enigmatic Denisovans. Surely, there is much more to come on this topic.

Discoveries of Cro-Magnon Reveal Our Direct Descendants

Clay and Ground Bone

The year was 1868. Railroad workers digging into a limestone cliff near the village of Les Eyzies in the Dordogne region of southwest France came across bones and tools. They notified local scientists who uncovered four human skeletons: a middle-aged man, a younger man, a young woman, and an infant. They also found flint tools and weapons and ornaments made of shells and animal teeth. The cave was named Cro-Magnon, after a local hermit named Magnon who had lived there. The term now generally refers to the modern humans who lived at various places around the Earth during the period 10,000 to 40,000 years ago. Most paleoanthropologists believe that the evolutionary line progressed from *Homo erectus* (the first *Homo* species to leave Africa) to Cro-Magnon, who was a primitive form of *Homo sapiens*. The paleoanthropologists who subscribe to that theory believe that the Neanderthals became extinct because they could not compete during the 10,000 or so years during which they overlapped with Cro-Magnon (30,000 to 40,000 years ago), and that Cro-Magnon is the *only* direct link back to *Homo erectus*. A second, similarly accepted theory states that *H. erectus* evolved to *H. sapiens* in Africa and populated the world via migration from 100,000 to 50,000 years ago. Under either theory, we are clearly the descendants of *Homo erectus* (Figure 14-8).

Though many of the Cro-Magnon people lived in caves, others built shelters. They developed fishing and hunting tools, such as nets, spears, and traps. They had impressive artistic ability, as seen in their cave art and sculptures made of stone, bone, and ivory. They sewed clothes, made footwear, and knew how to make fire. They had complicated spiritual ceremonies and rites for hunting, burials, and other customs and activities. At the site of Dolní Věstonice in the former Czechoslovakia, they built kilns as long as 27,000 years ago, to fire a mixture of clay and ground bone into ceramics— humankind's first example of combining and treating two dissimilar substances to make a new and different product. It was not until 15,000 years later in Japan that any other group learned to turn clay into pots. The lives and lifestyles of Cro-Magnon were very similar to that of Native North Americans, Australian Aborigines, and a number of African peoples. At any one time, the Cro-Magnon people numbered about 3 million. They set the stage for our emergence into modern civilization, as we entered the last minute of the Year of the Universe.

It took more than 5 million years for humans to evolve from the little primates known as *Sahelanthropus* and *Orrorin* to the primitive hominid identified recently as *Ardipithecus* (Figure 14-4A) and then on to *Australopithecus* with stone tools in central Africa to *Homo erectus* (Figure 14-8) and finally the modern form of *Homo sapiens*. But it is only in the last 30,000 to 40,000 years that our species has had the physical and intellectual attributes that we now regard as those of human beings. Cro-Magnon men and women of 30,000 or 40,000 years ago were as intelligent as *Homo sapiens* of today because their intellectual *capacity* equaled that of modern humans. If you could pluck an infant from a Cro-Magnon community of 30,000 years ago and raise him or her in your home, the child would be no different in life and school from a modern child. This would not be true of Lucy or a Neanderthal child. We have simply built on the knowledge and experience of each generation to achieve the technological advances and create the social institutions that characterize modern civilization.

Formation of Distinct Races Is a Result of Natural Selection

Homo erectus *and Humidity*

Unique physical traits develop *within* each species as a result of surrounding environmental pressures, in a process similar to reproductive isolation, which was discussed in Chapter 13. Millions of years after hominids had become reproductively isolated from other primates, differences among them began to emerge because of their geographic isolation. They adapted to those unique environmental circumstances as the species gradually changed into *Homo sapiens*. In the most likely scenario, small groups or bands of *Homo erectus* migrated to various places all around the globe, then lived in isolation from one another as they evolved toward modern man, fighting to survive in a hostile world. However, that period of isolation was not long enough to result in speciation—the creation of separate species. That is, humans of each major race can mate with individuals in any other race and produce fertile offspring. DNA profiles of all the races show that our ancestors came from Africa. Just what happened after that is not entirely clear because *Homo erectus* has been found in Eurasia and in Africa, and it is now clear that there are common genes among Africans, Europeans,

FIGURE 14-11

Probable Routes of Expansion of *Homo erectus*

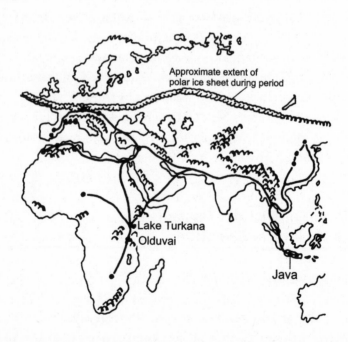

Asians, and Australians. That there are Neanderthal genes among us is the most recent and striking example of this complex science.

With the knowledge we now have about the gradually shifting crust of the Earth, the cycles of glaciation, and the effect of the environment on each species' evolution, we can understand how the geographically isolated groups of *Homo erectus* began to develop unique characteristics. For example, because overexposure to ultraviolet rays is harmful to skin, people who lived nearer to the equator formed a darker skin pigment than other people. The nineteenth-century German biologist and physiologist Carl Bergmann (1814–1865) proved that populations in the colder regions are bulkier than those in warmer areas because the icy wind caused their bodies to evolve as shorter and thicker to better preserve heat. Also, that environment caused their eyes to become squinted and protected by thicker eyelids. Bergmann's studies led to the principle known as Bergmann's Rule, which correlates external temperature and the ratio of body surface to weight in warm-blooded animals.

Other traits also represent valuable adaptations to the environment. For example, nose shape is related to humidity because moistening the air is a prime function of the nose. If the moisture content of the air is not brought to about 95 percent relative humidity as the person inhales, the lungs can be damaged. People in dry areas (such as Asians) tend to have narrow noses, the shape that causes the body to add more humidity while inhaling. Central Africans, on the other hand, have broad noses. Myriad other physiological traits—such as height, hair color and texture, heart rate, bone growth, fingerprints, and blood type—have also been studied and are attributable to environmental factors to which our ancestors adapted over a period of about 1 million years.

Modern scientists have been able to "watch" the process of natural selection in action in animals. For example, in November 1994, a biologist at the University of British Columbia reported on a population of stickleback fish that began to change shape and feeding habits in just a few generations when a new competitor forced the fish toward a different ecological niche. Countless studies of other animals, including twenty-first-century research on the finches of the Galapagos Islands, as detailed in Chapter 13, have provided objective observable data on this slow and silent process.

Geographical and climatic factors affect human evolution along with other external conditions, such as predation; disease; migration; conflict; and competition for breeding space, living space, and food. Over the million-year existence of *Homo erectus*, in response to all these stimuli, geographically isolated groups developed distinct gene pools and appearances that physical anthropologists have identified as races. Each of these groups developed its own language, culture, and history. But these are not distinct, nonoverlapping genetic groups of people. Regardless of race, we are all *Homo sapiens*, descended from *Homo erectus*. There clearly are genetic differences among human beings, but how the race concept is used is a matter of social convention. Whether the concept of race remains a meaningful and useful social convention is questionable as humans travel, intermarry, and blur the lines that separate human populations.

The Existence of Distinct Races Has Had Major Political Effects

The Door of the Gods

The simultaneous physiological and cultural development and differentiation of the various racial groups has now revealed itself as one of the most significant *political* facts in human history. About 10,000 years ago, humans changed from forager and hunter to animal raiser and farmer. Foragers and hunters traveled in bands averaging forty individuals, following herds of animals to ensure a food supply. But after the last Ice Age ended, there was a burst of new vegetation, and humans began to remain in one spot to domesticate animals and cultivate plants. We became shapers of the landscape. This was a significant change in the cultural and behavioral evolution of our species because it was then possible for people to create villages and communities and, indeed, create civilization as we conceive of it today. Before that change, it was difficult for technology to develop to any significant extent because the nomadic people had to carry everything with them on their daily journeys following the herd. In a culture limited to simple devices that had to be light, accessible every night, and packed away every morning, there was no room or time for creativity, innovation, experimentation, or expansion of the mind. Until humans settled down and created stable communities, the focus of daily life was survival. Our only ambitions were to follow the traditions of past generations and create future generations.

Until about 10,000 years ago, humankind's only inventions were the basic tools for hunting and fishing, and perhaps the kiln of Cro-Magnon. At about this time, copper smelting was developed and the wheel was invented in the Middle East or in the southern part of the former Soviet Union. Of course, the wheel contributed greatly to the onrush of technological development that began a few thousand years later, and eventually led to the use of rollers and pulleys in farming and for many other purposes. The invention of the wheel was followed by techniques for tanning hides, weaving, and making pottery. About 6,000 years ago, we learned how to combine copper and tin into bronze. Iron smelting was discovered about 3,500 years ago. By about the year 1000 BCE, steel was being made in India and its use spread to other parts of the world. The advent of permanent settlements and new technology had an extraordinary effect on world population, which increased from about 5 million to 86 million. This was more than a seventeen-fold increase in a period of 4,000 years—from 10,000 to 6,000 years ago. About 20,000 years ago, modern humans migrated east from Asia across the Bering Strait and remained for some time (perhaps thousands of years) at Beringia, a now submerged region in the strait between Russia and Alaska. These hunter-gatherers moved south, reaching Chile about 14,000 years ago. These were the immediate ancestors of the Paleo-Indians and Native North and South Americans. All human skeletal remains from North to South America are of modern *Homo sapiens*, and the genetic details show that all of these populations came from Asia.

By about 3000 BCE, humans had trained the ox and horse to draw carts, and by 2000 BCE, we discovered how to ride the horse and sail the seas. With boats, horses, wagons, roads, and bridges and with an expanding population and adventurous spirit, people began to cross mountain ranges, travel down rivers and into seas. They began to traverse the distances that separated once-isolated groups. As nature dictates, these groups competed. They fought over territory and food and for less important reasons. People spread out in cities and villages across the land, resulting in increasing contact with people who spoke a different language, who were lighter or darker, taller or shorter, believed in a different god, prayed to a different idol, dressed oddly, and, because of such differences, were often judged to be less human.

As competition increased, and territorialism evolved into nationalism, those differences have been used to justify oppression and war.

In the European provincial view of the world depicted in the literature of the Middle Ages, Europeans regarded physical deviations from the European "norm" as the result of degeneration or divine punishment for sin. Human history of most other cultures is replete with similar narrow-mindedness and self-aggrandizement. Cultures and tribes all over the Earth called themselves "the people" or "all men," and relegated other groups of humans to subhuman status. Mediterranean means the "middle of the Earth." For thousands of years China called itself the Middle Kingdom. In ancient Greece, Mount Olympus (abode of the gods) was the center of Greece, which was the center of the universe. The Hindus placed Mount Meru in the center of the universe, much like Earth was believed to be the center of the universe until Copernicanism finally took hold in the Renaissance. Babylon (meaning "door of the gods") represented the location where the gods came down to Earth. For the Moslems, their Kaaba was the highest point on Earth.

History, of course, is a complex subject. Conflicts throughout history result from a great variety of facts and circumstances at a given point in time. However, the physical and cultural differences among groups and nations undoubtedly have been one of the primary bases for such conflicts over the last several thousand years. This includes the conflicts among racial groups within the United States and other countries. The conflicts began with the earliest tribal battles and swept into the twentieth century with incredibly brutal force. Religious and cultural differences have been the root cause of more deaths in the nineteenth and twentieth centuries than in all prior centuries combined.

Because *Homo erectus* dispersed, then evolved into multiple races and cultures, world history overflows with the accounts of military conflicts, the domination and enslavement of peoples throughout the world, and outspoken hatred between physically and culturally diverse groups. If we had understood before now the real mechanisms and history of human evolution, would there have been less death and destruction? If people understood that their appearances and cultural differences are the result of evolution, and not divinely inspired or a reflection of innate superiority, perhaps they would not be so quick to condemn or so eager to torture, maim, and kill one another.

Posed another way, if *all* human beings had evolved at one central geographical location until relatively recently (perhaps within the last few thousand years), natural selection would not have had time to cause noticeable physical or cultural differences among us. So, for example, the result might have been that all human beings would now have black hair, blue eyes, brown skin, and more uniform height and features. Most important, being from one group, we would all speak the same language and share similar cultural and religious beliefs. Faces and personalities would distinguish one person from another, but there would be far fewer differences among us. In such a world, we would never have enjoyed the richness and fascination of our diverse cultures, and the course of human history would have been drastically altered. People might have found other reasons to discriminate and enslave in a monocultural world. But could we have avoided the thousands of wars among nations and the millions of deaths and injuries and enslaved populations that resulted from cultural and religious differences?

We will never know the answer to that question. Yet, we could live in greater harmony by understanding human evolution and now knowing we are one species. We are all descended directly or indirectly from *Homo erectus,* who migrated out of Africa. Indeed, as we will see in Part Six, our first common ancestor is the one-celled animal that first appeared on Earth on September 25 of the Year of the Universe.

It is ironic that once we had reached the point in our evolution and development when our thought processes were sophisticated enough to discover the laws of nature, natural selection became far less of an influence on our everyday lives and our long-term destinies. Since about 10,000 years ago, when we organized ourselves into cooperative communities, it has mattered less how fast any one of us could run. Humans now survive in a hostile environment solely because we can change the setting, not because we can adapt to it by natural selection. We inherited all our physical abilities and traits from our successful ancestral species. But now our species is defined almost solely by its ability to contemplate the future, form a self-image, and reshape the environment, for better or for worse—an incredible capacity that is due solely to the forces of natural selection that molded our brains into the most powerful force on this planet.

We no longer play by the rules of natural selection. We change our environment to suit us. When infectious organisms make us ill, we invent drugs to kill them. If we cannot see well, we put on glasses, and if we are too slow to catch a rabbit, we pick up a gun and shoot it. This behavior has allowed us to become increasingly weak, more prone to infection, and more dependent on our brain power for survival. How long we will persist as a species depends on whether we can develop a plan to successfully maintain our environment. While the other billions of living creatures on this Earth provide the raw materials, niches for survival, and the pressures on natural selection that maintain the natural balance in our world, we are the one organism that understands these complex processes, yet we are forever radically altering them.

Although we now understand the process of evolution in great depth, we continue to see emotion swirling around it. In an effort to support an irrational view of evolution, there are many people who claim it is still a theory. Indeed, in 1859 it was a theory. Today, however, it is an absolute and inescapable principle and fact. Calling evolution a theory, and ignoring today's incredibly solid and powerful edifice of information and knowledge, is nothing more than a simplistic word game, an issue of semantics. Calling it a theory is perhaps the greatest misnomer in society and has nothing to do with science or scientific debate. Very simply, as Huxley told Wilberforce in their debate in 1860, because humans are "highly endowed by nature, we don't do justice to our species and do not deserve this fantastic ability to learn and to reason if we introduce ridicule into a grave scientific discussion."

Chapter Fifteen
CONTINENTAL DRIFT

*The continents are adrift, ferried about on great fragments of the Earth's
shell. . . . The plates shudder and quake more than a million times a year.
"Solid Earth" is not as solid as we thought. It is energetic, dynamic, and
fundamentally restless. . . . It is continually creating its own surface,
destroying it, repairing it, renewing it like a skin.*
                                        —Jonathan Weiner, *Planet Earth* (1986)

Well before Darwin wrote *The Origin of Species*, people were aware of the
complex interrelationship of plant and animal life in ecosystems, and the
effect of climate, topography, and other environmental factors on survival.
With the advent of evolutionary theory, the view of that interrelationship
was expanded to include natural selection. But a new and revolutionary discovery in geology beginning in the 1960s presented a critical aspect of the
evolutionary process. At the same time that paleoanthropologists were
collecting the fossils that enabled us to trace our ancestry back more than
65 million years, new geologic evidence was emerging explaining a startling
and previously unknown cause for many of those external environmental
changes and pressures that trigger natural selection.

With the San Andreas Fault in California as the dividing line, two giant
land masses known as the Pacific Plate and North American Plate (two
of ten major plates of the Earth's crust and mantle that cover its surface)
grind and slide past each other at about the same rate as our fingernails
grow, about one and half inches per year. In about 15 million years, Los
Angeles and San Francisco will sit directly next to each other. Six major
fault systems underlie Los Angeles alone, and a hundred active faults pose a

constant threat to this major population center. Though there is virtually nothing that can be done to diminish this threat, we now understand its fundamental causes and realize that it is part of a global phenomenon that has had a major impact on all life, not only the citizens of California, for millions of years. The discovery that the plates exist and knowledge of plate tectonics—the dynamic processes of continental drift—has unified geology and evolution in a remarkable way.

Earthquakes and Volcanoes Give Rise to Mythology as They Cause Devastation and Death throughout the World

Whales, Boars, and Snakes

Lisbon (Portugal), 1755: Violent shaking kills 60,000 people, as they are drowned from the tidal waves produced at the earthquake's center off the coast, crushed by the 12,000 buildings destroyed, and burned in the fire that lasts six days after the initial shock.

Laki (Iceland), 1783: Ashes from this volcano ruin vegetation, leading to famine and epidemics resulting in the death of 10,000 people.

Unzen-dake (Japan), 1792: Mud streams caused by this volcano destroy villages on its slopes, killing 10,452.

New Madrid (Missouri), 1811: The first of a series of earthquakes and aftershocks that lasts for nearly two months are felt in a radius of a thousand miles, and cause widespread flooding and damage.

Tambora (Indonesia), 1815: More than 12,000 die from this volcano's shower of rock and ashes. The resulting famine causes the death of 80,000 people on the neighboring islands of Sumbowa and Lombok.

Krakatoa (Indonesia), 1883: Having been dormant since 1660, ash-laden clouds spew forth beginning on May 20, reaching a height of six miles. The volcano is active intermittently over the following months, until it climaxes on August 27 with perhaps the most violent eruption in modern history. The explosions are heard 2,200 miles away in Australia, ash is propelled 50 miles into the atmosphere, and the eruption triggers tidal waves that kill 36,000 people in nearby Java and Sumatra.

Mount Pelée (Martinique), 1902: A volcanic cloud of ash and lava kills 29,000 people as it sweeps over the town of Saint-Pierre.

Kelud (Java), 1902: When this volcano erupts, the mudflow and flood resulting from the release of waters from the crater lake causes the death of 5,100 people.

San Francisco, 1906: On April 18, at 5:12 a.m., the San Andreas Fault slips along a segment several hundred miles long, resulting in the death of 700 people and causing a fire that destroys the central business district of the city.

Tokyo, 1923: A violent earthquake causes an estimated 140,000 casualties, as nearly half of all the structures in the metropolitan area collapse.

Chilean Coast, 1960: Over a distance of 1,000 miles along the coast of Chile, 5,700 are killed and 3,000 are injured from this quake. The resulting tidal waves in Hawaii, Japan, and the Pacific coast of the United States cause further death and destruction.

Tangshan (China), 1976: On July 28, an earthquake near this coal-mining and industrial city east of Beijing causes the most devastating natural disaster in modern history—240,000 people are killed and 500,000 are injured.

Haiti, January 2010: A massive earthquake rocks the Caribbean island, killing more than 250,000 people. A resulting cholera epidemic remains a severe problem in 2012, having killed thousands more inhabitants.

Tohuku (Japan), 2011: On March 11, a magnitude 9.0 earthquake generates a tsunami that devastates a wide area of Japan north of Tokyo. It is the most powerful earthquake in recorded history to affect Japan, and the tsunami kills about 20,000 people, leaves millions homeless and produces a nuclear accident at the Fukushima Power Plant that remains a threat to local populations.

According to the mythology of a tribe in the Philippines, a goddess by the name of Aninito ad Chalom was the cause of all earthquakes. In the legends of the Bhuiya tribe of India, the goddess Baski Mata stood on her head at the bottom of the sea, supporting the Earth on her feet. Whenever she tired and shifted position, there was an earthquake. The primitive people of Chile believed a monstrous whale lived inside volcanoes. In India, the cause of volcanoes was a gigantic mole or ferocious boar. Indonesians believed that when the snake Hontobogo moved, the Earth shook and fire erupted from the mountains. Animals were sacrificed in Sumatra to appease the spirits that dwelled in volcanoes. The Bagobo tribe in the Philippines sacrificed humans. Myths about Cyclops and Titans were inspired by volcanic activity.

Aristotle believed that volcanic explosions and earthquakes were caused by hot winds that move underground and occasionally burst forth in volcanic activity and earthquake tremors. He believed that these hot winds were caused by compression that was created when waves breaking on the seashore forced such air into caves that exist along the coast. Like many of Aristotle's other theories, this "pneumatic theory" endured for over 2,000 years, only to be ultimately disproven.

In March 1996, the previously dormant volcano El Popo (for Popocatépetl, Aztec for "smoking mountain") forty miles southeast of Mexico City suddenly came to life and began belching ash and steam. Indians in the village of Xalitzintla, and many more among the 400,000 people in El Popo's shadow, trekked up the ash-coated slope of the 18,000-foot volcano with offerings of fruit, flowers, and chile sauce in the hope it would not erupt and bury them in lava. Primitive myth still flourishes as current reality in many cultures—to make the world intelligible and gain influence over it; to ward off disasters; to reduce fear; and to escape from powerlessness, insecurity, and confusion in the face of overwhelming physical phenomena such as earthquakes and volcanoes. Fear first created the gods—people feared that their village would be next, that it would be buried in ash or shaken to its foundations. They worried they would be taken without warning. In light of the history of death and destruction caused by earthquakes and volcanoes, it is no wonder that these once-mysterious activities generated so many attempts to explain them and became an integral part of cultures around the world.

Beginning with James Hutton and Charles Lyell, geologists through the nineteenth and into the twentieth centuries made inroads into understanding the causes of earthquakes and volcanoes. However, as the story unfolded in the mid-twentieth century, geologists began to realize that truth was stranger than the fictional mythologies that centered on these activities. Though Hutton and Lyell recognized a constantly changing Earth, they never imagined what modern geologists would eventually discover about the true cause of these phenomena.

Massive Plates of the Earth's Crust Gradually Redraw the Continents

Pangenesis and the Island of India

The theory of continental drift was originally proposed in 1912 by the German geophysicist and meteorologist Alfred Wegener (1880–1930). When he studied the similarity of the coastlines of eastern South America and western Africa, he speculated they had been joined together until about 250 million years ago as part of one giant continent called Pangaea (*pan* meaning "all," and *gaia*, meaning "Earth").

Contrary to other theories about shifting continents, Wegener's Pangenesis theory stated that the continents had broken off of Pangaea and slowly drifted thousands of miles apart, like pieces of a cracked ice floe. In his 1915 book, *The Origins of Continents and Oceans*, Wegener noted that the Appalachian Mountains appear to be part of the same line of mountains that goes up through Nova Scotia then down through Scotland and Scandinavia, that an east–west mountain range in Argentina matches one in South Africa, and that a plateau in Brazil fits into a similar structure in Africa's Ivory Coast. The examples he relied on were not limited to the physical similarities. Fern fossils at certain locations around the world also matched to such a remarkable degree that they alone could justify the Pangenesis theory. Despite his substantial evidence, by the time Wegener lost his life on an expedition in Greenland in 1930, most geologists discounted his theory of continental drift. Pangenesis sank into obscurity for decades. As late as the mid-1950s college geology textbooks reflected our lack of knowledge about mountains, volcanoes, and earthquakes. For example:

FIGURE 15-1

Pangaea

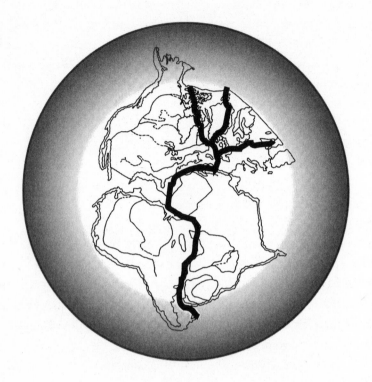

The cause of the . . . mountain ranges is not yet established. . . . Until a much more complete knowledge of . . . the interior of our planet is obtained . . . the origin of mountain ranges and volcanoes will remain unsolved. . . . The positive geological results attained by seismological methods . . . foreshadows a new and exceedingly interesting era of . . . research.

Indeed, the science of geology was on the verge of an enormous advance. As a result of extensive exploration of the seafloors to track submarines during World War II, combined with other geologic evidence that began to mount in the 1950s, a more detailed and accurate version of the theory of continental drift evolved into the modern field of plate tectonics. The evidence is now overwhelming. The new view of planet Earth is primarily the result of these discoveries:

- *Similarities in fossilized organisms* in strata of distant continents along areas where they would physically fit (such as South America and Africa)

- *Similarities of rocks and structures* of distant continents in those same areas

- *Consumption of the Earth's crust* in areas around the world, such as the northern edge of the Australian Plate being forced under the Eurasian Plate (forming the Himalayas), and the western edge of the Pacific Plate being forced under the Asian Plate (resulting in the proliferation of volcanoes and earthquakes in Japan and the surrounding area)

- *Configuration and composition of ocean floors*, such as the division between plates at the 46,000-mile-long Mid-Ocean Ridges of the Atlantic and Pacific, caused by the generation of new crust boiling out of the Earth's mantle (this phenomenon is known as seafloor spreading and is determined by age dating of minerals and sediments that make up the ocean floor)

- *Direction of rock magnetism in ocean floor* as shown by "striping," which exists in mirror-image fashion, tracing its "growth-lines" to expansion of ocean plates

The direction of magnetism of minerals in the ocean floor has provided fascinating and conclusive evidence of seafloor spreading. For many years we've been able to measure the intensity and direction of the Earth's magnetic field with instruments called magnetometers, towed behind ships. We also know that the polarity of this field changed naturally about 700,000 years ago, and that at the time rock is formed at the Earth's crust, including at the ocean floor, it takes on the same polarity as the geomagnetic field in which it exists. The measurements by magnetometers revealed 180-degree deviations in certain rocks that make up the ocean floor, particularly at the crests of the mid-ocean ridges. That is, intermittent portions or strips of sediment had the exact opposite polarity or magnetism from the strips next to them. In 1963, two British geophysicists, Frederick J. Vine and Drummond H. Matthews, proposed that the deviations in magnetism found in these sediments resulted from volcanic rock steadily emerging from beneath the Earth's crust and spreading in both directions as the Earth's magnetic field reversed itself. As they matched the date of such rock formation to the crust on either side of the great oceanic ridges, parallel strips formed, much like the matching rings of trees or like matching up rock strata at distant locations. The results of their research revealed that the polarity of the young igneous rock on the ocean floor matches from one side of the mid-ocean ridge to the other, as illustrated in Figure 15-2.

The pace and progression of seafloor spreading is one to five inches a year and is reflected in the exact bilateral symmetry of the magnetic bands on either side of the ridge. This information revolutionized concepts of the origin and age of the ocean floors and confirmed the phenomenon of seafloor spreading, which in turn was instrumental in resurrecting Wegener's original theory about Pangaea and continental drift. Alfred Wegener didn't have available to him the sophisticated equipment, detailed tests, satellite images, and computer modeling that can now explain the Earth's internal dynamics and trace its history. But after decades of refining global theory we now see how continents can actually "drift." Each plate is made of the Earth's crust and the top portion of the next layer of material, called the mantle. Each plate ranges from a few miles to as much as sixty miles in thickness, and "floats" on the underlying mantle, which is molten. As shown in Figure 15-3, the plates' edges have no particular correlation to the seven continents or to shorelines because the location of oceans and seas, of course, is due to land elevation, not the plates' shapes or boundaries. If we entered another ice age, glaciers and polar caps would enlarge, causing the sea level

FIGURE 15-2
Seafloor Spreading at the Mid-Ocean Ridge

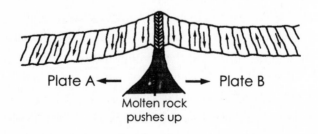

Plate A ← → Plate B

Molten rock
pushes up

FIGURE 15-3

Cross-Section of Plate/Continent

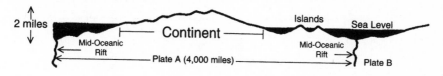

worldwide to drop and the portions of land above sea level would increase. If, on the other hand, the Earth's temperature were to increase over the next few thousand years, the polar caps would melt and the entire Earth would be covered by water. Movement of the plates themselves has also changed shorelines dramatically back and forth across the continents over the past tens of millions of years, apart from global changes in the ratio of ocean to ice. Thus the label *continent* is merely the identification of that 25 percent portion of the Earth's plates that happens to be above sea level at this particular point in the Earth's history. Continental drift is a misnomer because it is actually the *plates* that are drifting, with the highest portions being exposed to air as they ride about on those largely submerged plates in a dynamic process by which the Earth's surface is continuously redesigned.

At one edge of each plate, fresh molten rock flows up from the Earth's mantle and causes the plates to push apart. This is the seafloor spreading shown in Figure 15-2. At the other edge of the plate old rock is pushed down into deeper zones where it is melted down again, or collides directly against the opposing plate's edge. Fifty million years ago, India was an island, slowly moving toward Asia. Though India is now part of the Asian continent, it is actually the northern tip of the Australian Plate, which was pushed higher when it collided with the Eurasian Plate.

The Himalaya Mountain Range welled up along the line where these land masses crunched together. Over the last 40 million years, Mount Everest, composed of rock that once laid on the calm sea bottom, was thrust up to become the highest point on Earth. Similarly, in early 1996, researchers found rock high in the Swiss Alps that had once been 400 miles deep in the Earth's mantle and had been thrust upward as a result of continental collisions like the one now taking place between the Australian and Eurasian Plates. North America, like many other continents, is a conglomeration of

FIGURE 15-4

Migration of India into the Asian Continent

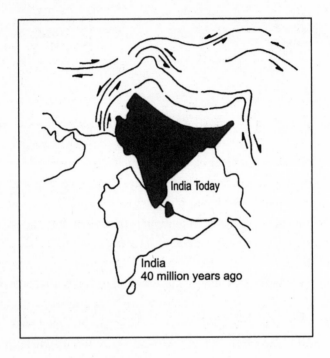

India Today

India
40 million years ago

pieces and blocks of plates that broke from larger plates and drifted together over the last 200 million years.

Earthquakes and volcanoes proliferate along the areas where the plates rub and chafe past each other in slow motion around the world.

The outline of most plates is evident from the number of earthquakes and volcanoes at their edges, as experienced firsthand by the residents of Japan, which is shaken a thousand times a year, California, and other places around the world that happen to straddle these massive and dangerous cracks in the Earth's surface. One million quakes or tremors rock the Earth's surface every year. About 50,000 earthquakes are measured with seismic instruments every year, and about a hundred of these are large enough to cause significant damage. Tokyo has been laid to ruin by earthquakes an average of once a century for the last 2,000 years. It is not surprising that Japan spends about $100 million each year on earthquake prediction research.

FIGURE 15-5

World Map of Plates, Identifying the Distribution of Earthquake Epicenters (Black Areas) Related to Plate Boundaries

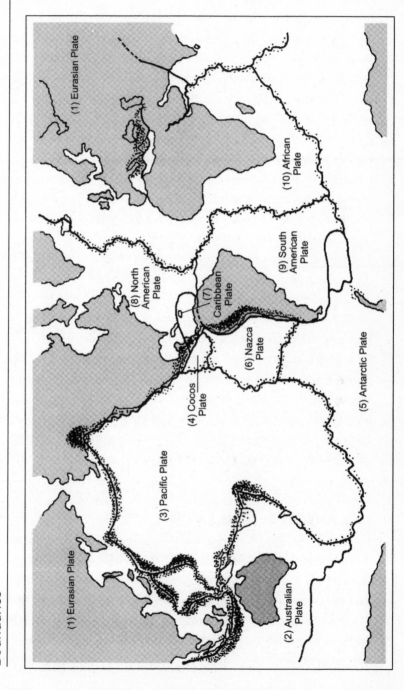

Ironically, Japan owes its very existence to volcanoes caused by the same series of phenomena that cause the earthquakes. The East Pacific Rise (the counterpart of the Mid-Atlantic Ridge) is the result of seafloor spreading that pushes the Pacific Plate toward Asia. However, as the Pacific Plate advances, it is sliding under the Eurasian Plate and into the Earth's interior, resulting in tremendous friction that causes some of the most violent earthquakes and volcanoes on Earth. The islands that make up Japan have risen out of the sea over the last 30 million years as the direct result of the volcanic activity at this juncture of plates. As recently as April 1973, volcanic eruptions south of Tokyo caused lava to rise above sea level to form the new island of Nishino-shima. Similarly, the underwater volcano Loihi (known as a seamount), twenty miles southeast of the island of Hawaii, last erupted in 1996, remains continuously volcanically active, and its peak—which is now 3,000 feet below the ocean surface—will form the next Hawaiian island in about 100,000 years.

The forces driving this ponderous movement originate in the Earth's core and mantle. The iron core is 1,500 miles in diameter, has a pressure 3.6 million times the pressure at the Earth's surface, and has a temperature similar to that of the sun's surface at a mere 10,400°F. As this tremendous heat is released, it causes convection and massive disruptions in the Earth's mantle. The mantle, the dense layer immediately below the Earth's lighter crust, is about 1,800 miles thick. The rigid, topmost 40 miles of mantle, together with the crust, form the tectonic plates. As one goes deeper, the layers of the mantle are being heated by the core, and churn like thick soup in slow motion. The earthquakes and volcanoes are the superficial results of the grand motion and rise of heat from the insides of our planet.

Life Adapts as Plates Drift

Two Months without Food . . . Life in Total Darkness

As these massive plates slowly dance with each other, all of life is affected. There are unmistakable remnants of tropical jungles in Alaska, and the marks of great glaciers beneath the Sahara. This massive African desert was covered by lakes and forests only 130,000 years ago. The plant and animal life that propagated around the Earth on the continents and in the seas has been riding the plates for hundreds of millions of years. The plates carried life with them from one location on this sphere to another, like a raft slowly moving downstream. Some plates began in a cold and dry region and ended up in a tropical environment, whereas others have applied their tremendous force against each other, resulting in the crust buckling into the world's massive mountain ranges. Plants and animals adapted as these rafts made their way along the route. The land beneath them rose and fell over millions of years, as natural selection dictated the physical and intellectual characteristics favorable to survival of each species. Thus plate tectonics has profoundly affected the history of life on Earth.

Great spans of time are prerequisites for such adaptation. This is seen in what could be called the "Iron Turtle Competition"—the remarkable history of the green turtles that migrate 2,000 miles from Brazil to tiny Ascension Island in the middle of the Atlantic Ocean every year. Why don't they just stay in Brazil and nest on the hundreds of miles of beaches there? What explains the choice or "programming" of these 400-pound reptiles to undergo this seemingly impossible trip—without feeding and in the face of myriad dangers? How do they even know the island is out there? Only one explanation makes good sense. It is based primarily on the fact that tens of millions of years ago there were islands only a stone's throw from South America. The ancestors of these turtles made the short swim to nest and lay their eggs at a location where the eggs and young would be secure from the many predators on the mainland. As the seafloor spread and Pangaea separated into Africa and South America, the ocean expanse gradually increased between Brazil and the series of volcanic islands that have come and gone in the vicinity of Ascension. From one generation to the next, the turtles unconsciously adapted to the increase of a few inches a year, caused by the lava welling up between the plates. But as they adhered to a migration pattern that seemed identical to the tradition of their ancestors, they were actually committing their progeny to this incredible task millions of years later. Thus the green turtle evolved into a larger animal and a stronger swimmer in order to survive and produce offspring. It developed heavy fat deposits and a metabolism that would enable it to travel 2,000 miles over a two-month period *without food!*

As mentioned earlier, lava erupts along portions of the mid-ocean ridges.

A variety of plant and animal life has evolved in these deep undersea "oases," just discovered in 1973 by scientists in small submarines researching the rift valley at the center of the Mid-Atlantic Ridge. The giant mussels, clams, crabs, fish, giant tube worms, and unique plant life flourish there and yet totally depend on the heat and minerals being spewed forth by tectonic activity. A unique "pocket" of natural selection, these vents that originated in the mantle are the only places on Earth where life is not directly or indirectly a product of sunlight. Instead, the plant and animal life at these depths, where there is no sunlight, consume bacteria that derive energy from hydrogen sulfide that results when the water mixes with the molten basalt beneath the ocean floor. Thus, in a relationship even more direct than the one experienced by the powerful green turtles, the creatures that live along these undersea rifts owe their very existence to plate tectonics.

Plate tectonics has also been a direct cause of reproductive isolation (discussed in Chapter 13) ever since Pangaea began to break apart. We've already seen numerous examples of reproductive isolation—from the differences in the finches' beaks and tortoiseshells in the Galapagos Islands to the unique physical traits among the races of humans that used to be geographically separated. We continue to discover previously unknown effects of plate tectonics. In the fall of 1995, a team of French and British scientists on an expedition deep into the unexplored Riwoche mountain region of northeast Tibet entered a seventeen-mile-long icy valley, isolated by 16,000-foot-high passes on both ends. They discovered an entire herd of an archaic-looking horse that had never been seen or known before by biologists. The direct result of reproductive isolation caused by the formation of the Himalaya Mountains, this 4-foot tall animal had branched off over 5 million years ago. It has retained several unique and primitive physical characteristics that distinguish it from all other breeds of horse, such as its small stature and triangular head shape that resemble the images of the vanished horses seen in Cro-Magnon cave paintings.

In 2009, Dennis McCarthy asked in the book *Here Be Dragons* why the various animal species are distributed where they are across the world. The field he has embraced to answer this complex question is called "biogeography." The "dragons" to which he refers are the largest living lizards on Earth today, the Komodo dragons. They inhabit several Indonesian Islands on which these giant reptiles obviously were reproductively isolated and evolved to clearly dominate their limited ecosystems. Using such striking examples, McCarthy reminds us of the central role that plate tectonics has played in the evolution of plants and animals worldwide. He brings us back to Darwin's compatriot Alfred Russell Wallace, who was the first to recognize the distributional differences between species in Australia and Asia. It was Wallace who actually originated the term "survival of the fittest," and it is the "Wallace Line" running between the South Pacific Islands of Borneo and Sulawesi that gives us such a clear perspective on how plate tectonics isolated biological populations and spurred their evolution. While collecting specimens on Ternate, Indonesia, not far east of the Wallace Line, he wrote the paper titled "On the Tendency of Varieties to Depart Indefinitely from the Original Type." At the same time this paper was presented in London in 1858, Darwin presented his notes on this topic. But it was Wallace who unknowingly had defined a portion of the Pacific and Australian tectonic plates that now hold such very different (but related) flora and fauna. Wallace, a contemporary of Charles Darwin, was the only other person in history who independently imagined the earthshaking concept of survival of the most fit individuals as the driving force in evolution. In 2012, all 28,000 pages of his writings, as well as associated documents and images, were compiled in a fascinating website (wallace-online.org). Some consider Wallace the cofounder of the principles that explain evolution. Russell clearly recognized this truth, but it was Darwin who published the original book, *On the Origin of Species*.

One sweeping global theory, plate tectonics, has fused broad perspectives and diverse information discovered by geologists, biologists, and physicists. These scientists have now accounted for the mid-ocean ridges, the world's great mountain ranges, almost all of this planet's earthquakes and volcanoes, the birth and disappearance of oceans, the movement of continents, and an essential chapter in our knowledge of evolution.

PART SIX

The Cell and Genetics

In Focus

Eleven billion years after the Big Bang a handful of elements combined on this Earth to begin an incredible process. Life emerged from the primordial soup. Life from lifeless electrical charges. Life that can now be explained in terms of physical constituents and processes, even though such explanations might seem too cold and dispassionate to account for the richness and complexity of living things and human consciousness.

The origin of life on Earth, reproduction of living things, and Darwin's theory of evolution were stripped down to their physical and chemical basis in the twentieth century—down to the point at which we were able to observe the previously undetectable "factors" of inheritance, the "mainspring" and mechanism of evolution. For the first time, we could understand the interrelationship of those three elements of life: origin, reproduction, and evolution.

First, we must consider the cell and genetics examined in this Part Six, then we look at the blueprint and backbone of life in Part Seven. From the cosmic egg in Part Four we now explore the human egg. From the cosmos to the chromosome. From the stark and bare expanse of Hubble's galaxies, to the infinitesimal, the microscopic, the molecular, the unseeable, the unimaginable within that is sustaining us. From the physicists' atomic energy to the biophysicist's energy of life. From evolution's macro view of life to the biochemist's micro view.

The story of cells, genes, and the molecule of life in Parts Six and Seven explains how thousands of people united the knowledge accumulated in diverse fields into one cohesive resolution, beginning with the discovery of cells in 1673 and the meticulous labors of an obscure monk at his monastery in what is now Slovakia in the 1860s.

Chapter Sixteen
PRIMORDIAL SOUP

Biology . . . is primarily a descriptive science, more like geography, dealing with the structure and working of a number of peculiarly organized entities, at a particular moment of time on a particular planet.
—John Desmond Bernal, *Science in History* (1965)

Darwin had accumulated volumes of data and examples to support his theory of evolution. Evidence and observations accumulated by scientists in the succeeding decades gave more support to the validity of evolutionary theory. Yet despite all this, the theory lacked *evidence from within* the living organism. Thus Darwin's theory of evolution begged the question: *What is the mechanism by which this natural selection works?* Exactly what is happening inside these living organisms to allow evolution to take place?

Well before *On the Origin of Species*, biologists and others also attempted to understand reproduction of living things and what directs, determines, and controls their growth, as we saw in Chapter 12 with the theories of spontaneous generation and preformation. The discoveries made in biology in the nineteenth century, together with evolutionary theory, set the stage for unraveling the mystery held deep within us and answering the most fundamental questions of life: *How did life first arise on Earth? How do organisms reproduce? What is the internal physical basis for evolution?*

The Cell, Its Nucleus, and Its Process of Dividing Are Discovered

Animalcules and Crucial Signals

What the atom is to physics, the cell is to biology. The atom is the building block of all matter. The cell, the building block of all life. Despite the incredible diversity of living things in this world, there is an underlying unity. All cells work essentially the same way, and *the genetic code that governs all life forms is identical in all species—conclusive evidence that all forms of life share a common ancestor*, as Darwin had predicted in *The Descent of Man* in 1871, well before anyone understood the role of the cell in heredity or reproduction. Now we have discovered that each cell is a tiny chemical factory that is capable of processing its own nutrients, generating its own energy through the use of those nutrients, communicating with neighboring cells, and *dividing* into two identical units. *The cell's ability to replicate itself is the key to all life and growth.* A single cell can be a complete organism in itself, such as a bacterium or amoeba, while other cells are specialized as the building blocks of multicellular organisms.

Though the average size cell contains trillions of atoms, it is still much too small to be seen with the naked eye—it would take about 10,000 human cells to cover the head of a pin. So, no one imagined the existence of cells until they could be observed under the microscope. In his 1665 publication, *Micrographia* (Small Drawings), the English physicist Robert Hooke (1635–1703) coined the word *cell* before any actual living cell was seen. He used the term to describe dead plant tissue and the microscopic honeycomb cavities in cork because the cellulose walls of the cork reminded him of the blocks of small rooms or "cells" occupied by monks in a monastery. Single lens magnifiers were first developed in the mid-fifteenth century but weren't powerful enough to see *living* cells of animals until van Leeuwenhoek produced more powerful lenses in the 1670s. In 1673, van Leeuwenhoek opened a whole new world when he discovered blood cells (adopting Hooke's word), spermatozoa, and single-celled organisms such as bacteria and swimming protozoans. Much like Galileo's first observations of stars in 1609, van Leewenhoek's initial glimpse of these cells and one-celled organisms was merely a small step in a journey that would unfold over the next three centuries. Neither man had any idea how immense and significant that journey would be, only that they could see objects that were not visible with the naked eye and that were unimaginable before those observations.

Much like seventeenth-century astronomers' efforts to label and mark the location of the newly discovered stars, and similar to Linnaeus's system of classifying animals in the eighteenth century, van Leeuwenhoek's discovery of what he termed lively "animalcules" first led to categorization of one-celled animals. In subsequent years, doctors and biologists categorized various types of tissues by viewing their cells under the microscope. For example, cells that make mucus were found to make up the linings of the airways and gastrointestinal tract, whereas those making fibrous connective tissue constitute the layers under our skin and surrounding our muscles. Three different kinds of muscle cells were discovered (cardiac, involuntary, and voluntary muscles) as well as numerous cell types specialized as components of the nervous and reproductive organs. However, even as this labeling effort progressed rapidly through the eighteenth century, the function and contents of cells remained a mystery.

Because the early microscopes were not strong enough to view all the details within cells, and because of the persistence of some of the archaic theories of spontaneous generation and preformation, cell theory didn't begin to develop until 1831. In that year, Robert Brown (1773–1858), a Scottish botanist, observed the control point of the cell, calling it the "nucleus," and identified that structure as being a common element of all plant cells—a discovery that rivals the importance of the later discovery of the atomic nucleus. Nuclei were soon discovered in animal cells, and the flowing of "protoplasm" (the internal material in the cell, a term that is no longer used) was observed in living cells in 1835. Two German biologists, Matthias J. Schleiden (1804–1881) and Theodor Schwann (1810–1882), refined and developed this early information into the first basic principles of cell theory. In 1838 and 1839 they wrote that all cells are composed of an outside limiting membrane, nucleus, and cell body and are the elementary particles from which all plants and animals are constructed.

Many theories about the "birth" of new cells developed during the 1840s but were discarded in favor of cell division, the process by which two identical "daughter" cells result from the original. "All cells come from cells," asserted the famous German physician and biologist Rudolf Virchow (1821–1902). In his 1858 book, *Cellular Pathology*, Virchow explained that life is not a result of supernatural phenomena and that "cells are the link in the great chain of . . . formations that form tissues, organs, systems and the individual." This established the modern concept of cell pathology—that is, the study of disease processes beginning at the cellular level. After Virchow and others proved the importance of cell division, cells became the subject of intense research, soon leading to the general acceptance of the fact that living things can arise only from the seeds of their parents. Life is never created anew on Earth but is descended from ancestors in an unbroken line. Coincidentally, Virchow's book was published in 1858, the same year as the joint paper by Darwin and Alfred Russell Wallace was read to the Linnean Society, and a year before publication of *On the Origin of Species*. However, it would take generations before evolution and cell theory would converge in a meaningful sense.

Biologists Discover the Cells' Organs

The Cell's Brain, Stomach, and Skeleton

Over the succeeding decades scientists began to identify other parts of the cell's internal machinery. Cells actually contain a set of distinct organs called organelles, not one "protoplasm" substance as once thought. Organelles are *ultramicroscopic* (that is, they can be seen in detail only with an electron microscope) and are internal compartments of various shapes, sizes, and complexities, containing sets of specific chemicals. Organelles carry out reactions on materials such as nutrients, growth factors, and toxic agents entering or leaving the cells. Cells are divided into three main compartments: the nucleus, the cytoplasm, and the cell membrane. If a cellular structure is not found inside the nucleus, it is in the cytoplasm or in the membrane that surrounds and contains the cytoplasm. The nucleus has a specialized limiting membrane as well.

FIGURE 16-1

Organelles

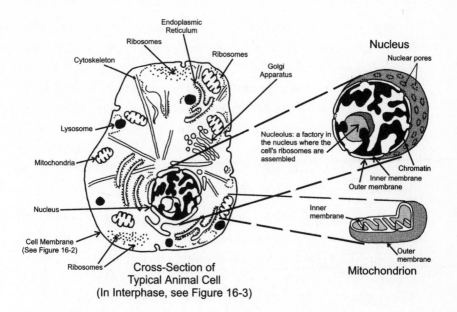

Cross-Section of
Typical Animal Cell
(In Interphase, see Figure 16-3)

Mitochondrion

- *Nucleus*: Taking up about one-tenth of the cell's volume, the nucleus is the cell's control center. It is surrounded by a complex membrane controlling the materials that go in and out of the nucleus and contains the nucleic acids (DNA and RNA) that constitute the genes that determine the composition of the proteins that control all the chemical processes in the entire organism (see how this works later in the chapter). RNA molecules, which carry the genes' instructions, are moved out of the nucleus (through its membrane) to the cell's protein-making organelle, known as the ribosome.

- *Ribosome*: These number in the thousands and sometimes millions in *each* cell, depending on how much protein the cell must make. Ribosomes are the attachment sites for the specific "messenger" RNA sequences that determine the proteins made by specific cell types.

- *Endoplasmic reticulum*: The endoplasmic reticulum (ER) is a floppy bag of folded membranes that takes up about half of the cell's volume. Its function is to provide a separate chamber in which the proteins made on the ribosomes can be put in their final and correctly folded form. This is where sugars, phosphates, and other molecules are added to the proteins at specific chemical binding sites.

- *Golgi apparatus*: A typical cell has thirty or forty of these smooth membranous structures. It is the only organelle named after a person— Italian biologist Camillo Golgi (1844–1926) who discovered it. It stores and processes fats and performs even further modifications of proteins as well as preparing them for transport to other parts of the cell or out of the cell entirely. For example, proteins that are digestive enzymes made in the liver need to be shipped to the stomach. In the Golgi, specific sugars attached to these digestive protein molecules provide the information that directs the protein to its final destination.

- *Lysosome*: Using powerful enzymes, these organelles form the cell's multiple "stomachs" or digestive sacks. These enzymes break down the food we eat into its smallest components, which then are used in the various cells. For example, lysosomes break down the proteins we ingest into amino acids that are reused by the cell to make new proteins. Enzymes contained in the lysosomes of white blood cells that protect us from infection destroy bacteria and parasites that are picked up by the cells.

- *Cytoskeleton*: These networks of ultrathin protein filaments maintain the cell's structure and shape, allow it to move, and support the cell's internal transportation network that links organelles and ferries products between them. The thinnest components of the cytoskeleton are called actin filaments. These highly active, very specialized proteins constitute what are called "molecular motors," providing the contractile forces that allow the cells to migrate and change shape (see Figure 16-3). Slightly thicker microtubules provide intracellular tracks and a substrate for the movement of the other organelles that are pulled by actin filaments throughout the cytoplasm.

- *Mitochondrion*: These sausage-shaped structures of folded membranes extract the energy from the chemical bonds of carbohydrates, fats, and proteins and convert that energy into a form that can be used by the other organelles to replenish the energy they expend in conducting their work. Amazingly, mitochondria have their own DNA and their own ribosomes, and they divide *within* cells. These characteristics, together with the fact they are shaped like bacteria and have bacterial DNA, indicate that mitochondria were likely once bacteria that originally entered cells as parasites and later evolved into this unique organelle.

- *Membrane*: The cell membrane (also called outer plasma membrane) not only holds all the cell's organelles inside but precisely controls the entry and exit of molecules that are critical to cell function. The molecules that make up the membrane are lined up unerringly with a "head" that attracts other heads and is compatible with the watery environment in which cells reside, and a "tail" that repels heads and water. This configuration allows for proteins to be embedded within or to span the entire membrane. Some of these proteins are actually channels that control what the cell takes in and sends out, whereas others form receptors that detect specific external signals, allowing the cells to respond to stimuli and conditions.

FIGURE 16-2
Cell Membrane Structure

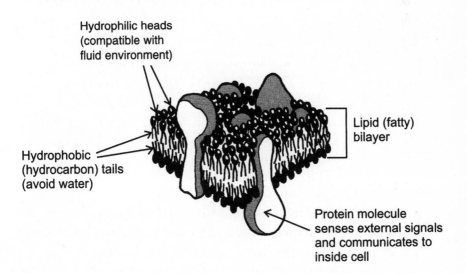

Hydrophilic heads
(compatible with
fluid environment)

Lipid (fatty)
bilayer

Hydrophobic
(hydrocarbon) tails
(avoid water)

Protein molecule
senses external signals
and communicates to
inside cell

Each cell contains only one nucleus, while the other organelles are present in multiple units. In addition to the organelles just listed, plant cells contain chloroplasts, which are much like mitochondria, but create energy through photosynthesis (that is, energy produced from light).

Biologists Determine the Phases of Cell Division

Dying of Old Age

Cell research reached an important phase with Walther Flemming's work in the 1870s. Flemming (1843–1905), a German physician and anatomist, improved the dyes used to see cell structures. In 1879, he was able to identify a threadlike material in the nucleus of cells. Observing that material (later called chromosomes) during cell division, Flemming showed that the threads shortened (that is, contracted and condensed) and split longitudinally in halves, with each half moving into opposite sides of two new identical cells. He called this cell-division process "mitosis" and described it in his 1882 book, *Cell-Substance, Nucleus, and Cell-Division*. He thereby founded cytogenetics, the study of the physical processes involved in inheritance at the cellular level.

After Flemming's work we gradually began to understand the physical processes within the cell. Yet, it took until the middle of the twentieth century to figure out what is contained in the cell and its nucleus that enables the cell to control its own functions and the functions of an entire organism. The "master plan" by which these cells undertake the complex steps to create a new single cell that proceeds to divide into 2, 4, 8, and ultimately 60 trillion cells that make up a complex living organism such as a human will be examined in Part Seven. Like everything else in the world, the physical process of each cell works according to the established laws of chemistry and physics. This includes the two basic steps of the reproductive process—cell growth and cell division (mitosis).

A major proportion of the cell's machinery is dedicated to the process of mitosis. This is because multicelled plants and animals must be able to replace cells that die naturally or are injured, and single-celled organisms

FIGURE 16-3

A. Stages of Mitosis

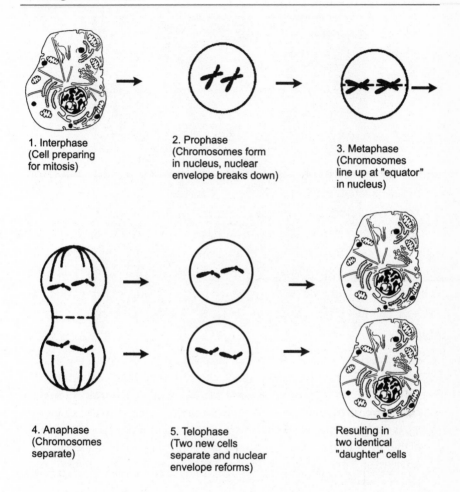

1. Interphase
(Cell preparing
for mitosis)

2. Prophase
(Chromosomes form
in nucleus, nuclear
envelope breaks down)

3. Metaphase
(Chromosomes
line up at "equator"
in nucleus)

4. Anaphase
(Chromosomes
separate)

5. Telophase
(Two new cells
separate and nuclear
envelope reforms)

Resulting in
two identical
"daughter" cells

must reproduce themselves. Regardless of whether we are considering a liver or lung cell of the human body, a plant cell, or a free-swimming protozoan, the process of cell division is essentially the same because the ultimate goal of the process is to produce new cells exactly like the ones from which they originated. Mitosis consists of the following five stages:

B. Mitosis

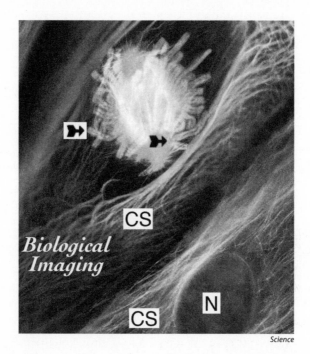

Science

In a special issue of *Science* focused on biological imaging, scientists captured a dividing cell in the metaphase stage of mitosis. Here the DNA has condensed into individual chromosomes (*arrows*) that are replicating so that the two closely apposed strands of new DNA can be seen. The linear strands surrounding the chromosomes and the nucleus (*N*) of the nondividing cell at the bottom of the image are the cables of actin and the microtubules of the cytoskeleton (*CS*) that control cell shape and migration and will guide the chromosomes into two new daughter cells, as shown in panel A.

1. Interphase: The cell spends most of its life in this phase. It is not replicating during interphase but is carrying out its essential functions, including making specific proteins, secreting lipids, and manufacturing the connective tissue matrix that holds us together. The pace of these activities depends on what the body needs at a given point in time. For example, every time we eat a meal, the cells that line our stomach and intestines must manufacture large amounts of proteinaceous enzymes and acids to digest the food. The cells making these materials are in the interphase of the cell cycle, but they are also fully dedicated to this secretory function and will live only weeks to months, soon to be released into the gastrointestinal tract at the end of their useful lives. Cells receive the signals to divide during interphase.

2. Prophase: A cell enters prophase after receiving an external stimulus to divide while in the interphase. During prophase, the long coils of DNA that are wound diffusely through the cell's nucleus begin to condense into organized structures called chromosomes. Thus the cell is preparing to divide. Its neighboring cell may have died of old age or other cells in the population may be under attack by toxic agents or infectious organisms. Whatever the reason, the goal is to produce two new identical cells. To do so, the chromosomes of that cell (thus the DNA containing the vital genetic information) must be replicated. The duplication of DNA—the result of a complex set of biochemical reactions—will be examined in Part Seven.

3. Metaphase: The defining characteristic of metaphase is the ordered lining up of all the chromosomes across the central "equator" of the cell (Figure 16-3B) so they are in a position to move equally to one new cell or the other. This critical event is controlled by precisely laid "cables" that will guide the chromosomes to their respective cells. The cables are part of the cell's cytoskeleton (described earlier). The nuclear membrane that was so prominent during interphase and much of prophase dissolves and no longer defines the nucleus. The entire function of the cell is now dedicated to a successful division into two new cells. Once metaphase begins, the cell division process (including the final two stages) typically lasts only a few minutes.

4. Anaphase: This marks the beginning of chromosomal separation, with the newly copied chromosome and its original having separated and moved apart. This chromosomal movement is mediated by the actin filaments of the cytoskeleton that are actually attached to the chromosomes and contract, thus drawing the new chromosomes into the new daughter cells. At the same time, microtubules spanning the two new cells provide tracks for correct chromosomal migration (Figure 16-3B).

5. Telophase: The chromosomes have now reached their respective cells and the two new cells are forming a new plasma membrane between them so they can perform their functions and reside side by side, such as in human tissues or, for example, so they can swim away as two new protozoans. The pairs of chromosomes soon become enclosed in a new nuclear membrane, and the cells move rapidly toward interphase, beginning the life cycle of the cell once more.

Thus the actual splitting of the cell is the final event of reproducing itself. What causes each phase? As a result of discoveries made primarily during the 1970s and 1980s, molecular biologists now understand the crucial events that trigger cell growth and cell division to an amazing level of detail. Complex chemical processes control each phase, ultimately directed by the DNA molecule, which contains the master plan of all life. But before looking at how 60 trillion cells, such as those making up one person, can possibly function as a single organism, we will look at one cell—the origin of the very first living cell on Earth.

Life Springs from the Primordial Soup

Vents and Tube Worms

There are no seas on this planet because it is so hot water would vaporize. Most of its surface is molten rock—the crust that does exist is tenuous, brittle, and fleeting. Meteorites plunge through a thin atmosphere of hydrogen and helium into a surface mix of methane, ammonia, hydrogen, carbon monoxide, carbon dioxide, and nitrogen. Along with the meteorites, intense ultraviolet radiation from the sun penetrates to such a high degree that a lethal dose for any living organism is delivered in less than an hour. The gravitational pull of this planet's moon—only 11,000 miles away—raises massive tides of molten rock.

This is not Jupiter or a planet in some other solar system. This is Earth 4.6 billion years ago, soon after it coalesced from dust and gases left over from the Big Bang into a shimmering sphere of molten material. This boiling, radiating inferno seems an unlikely location for the beginning of life. But over the next several hundred million years, the bombardment of meteorites slowed and the tremendous heat radiated out into space, allowing the surface to cool and stabilize. An atmosphere of heavier elements began to form, resulting in the formation of a layer of ozone in the upper atmosphere, which blocked the sun's harmful ultraviolet rays.

The first billion years of the Earth's existence are called the Hadean Eon, named after Hades, Greek for "hell." As stated by Jonathan Weiner in *Planet Earth*, "Life arose almost as soon as the planet ascended from hell." The first traces of life in the geological record are about 4 billion years old: simple one-celled organisms. What took place in the primordial soup to create what we now call "life"? As we will see in Part Seven, all life is made of complex molecules called amino acids, the building blocks out of which all plants and animals formulate their proteins. We will also see that molecules called ribonucleic acids (RNA) are essential for the existence of life because they "organize" those amino acids into proteins based on the genetic code fixed in DNA. However, amino acids and RNA molecules are themselves made by living things. That is, they are the product of complex workings within our cells, thus creating a chicken-or-egg riddle. How could life have developed from amino acids and RNA if those substances are the *products* of living cells? The answer lies in the probable existence of both of those essential components in the primordial soup of 4 billion years ago.

The basic chemical elements that make up amino acids and RNA were present at that time. In 1953, Stanley Miller, a graduate student at the University of Chicago in the lab of the famous chemist and Nobel laureate Harold Urey (1893–1981) (who played a major role in the Manhattan Project), designed a series of large interconnected glass flasks intended to represent conditions on early Earth. In Miller's simulation, steam rose from boiling water and mingled with methane, ammonia, hydrogen, and water vapor, the major gases present in this planet's early atmosphere. The mixture was then subjected to 60,000 volts of electricity to simulate lightning—all for the purpose of determining whether the critical amino acids could be created out of "whole cloth." After a week, Miller analyzed the chemicals in the water and found that it was rich with a variety of amino acids. This experiment has been reproduced countless times since 1953, always resulting in the formation of amino acids, the building blocks of proteins. Given the immense period of time and the billions of random combinations of chemical elements that must have taken place all over primitive Earth, these laboratory experiments provide a plausible explanation for the formation of amino acids, one of the molecules essential for life.

Likewise, scientists have demonstrated that the chemical constituents of RNA were abundant at the same time as amino acids were being formed. These basic elements (including phosphorus, nitrogen, hydrogen, and carbon) connected with one another to form long molecular chains by following the same physical principles of binding and valency that are followed by simple molecules. Certainly, some ribonucleic acids joined into chemical sequences that were meaningless. The mere existence of amino acids and RNA molecules does not explain how these building blocks eventually were configured and organized into living organisms. In order to be "alive," RNA needed two capabilities: the ability to organize the amino acids into proteins and the ability to duplicate itself so it could continue to exist and cause that essential process to take place. As for the first capability, RNA molecules most likely lined up randomly in a configuration that caused amino acids to form primitive proteins. These proteins could have promoted chemical reactions that would favor one RNA configuration over another; in other words, a form of "natural selection" of RNA sequences that are best at replicating and surviving in a hostile environment. RNA's ability to promote such chemical reactions was demonstrated in the laboratory in 1982 by the biochemists and Nobel laureates Thomas R. Cech (1947–) and Sidney Altman (1939–). This "self-catalyzing" capability was an essential step in the formation of life. It resulted in evolution from a strictly "RNA world" (as coined by Cech and Altman) to one in which RNA directed the manufacture of proteins. As we will examine in Part Seven, the synthesis of proteins by DNA and RNA is the basis for all life.

Regarding the ability of RNA to self-replicate, even the simplest and most primitive form of RNA had the potential for its own duplication through the formation of an identical complementary RNA molecule due to the principles of chemical binding mentioned earlier. *Once formed, primitive RNA began the first version of natural selection.* If various forms of RNA had existed simultaneously, the ones that could multiply most rapidly would have had the best chance of proliferating and further monopolizing the essential chemicals needed for forming such molecules. The RNA molecules that accelerated favorable biological reactions survived. In this system, evolution of RNA took place. Paraphrasing Darwin, natural selection worked solely by and for the good of each *RNA molecule*. Over millions of years, RNA molecules evolved into distinct and more complex types, each with a specialization, but all relating to each other in an organized and systematic manner. Thus RNA—the maker of proteins from amino acids—also had the consistent, overwhelming, and defining function of all living organisms— namely, self-replication.

Some biologists believe that the undersea vents at plate junctures (discussed in Chapter 15) were likely locations for the creation of life. At these depths, life was protected from the harmful ultraviolet rays. Also, essential elements and compounds like carbon, hydrogen, methane, water, and ammonia were present at these sites as well as a wide range of temperatures. With such a large number of locations throughout the world, like thousands of experimental labs, there was ample opportunity for the right chemical mix.

The formation of *membranes* around distinct groups of RNA and proteins assembled from amino acids must have been the next step toward cell development. As shown earlier, the lipid molecules that make up the cell's membrane naturally line up in rows that form an interior and exterior sheet. This compartmentalization created the first cells and allowed them to exist in a variety of watery environments. Like the primitive RNA molecules, these first cell "entities" competed for common ingredients, leading to variations in cells and the survival of those most efficient in passing on to their progeny the catalysts and other characteristics that promoted duplication.

Bacteria Become the First Common Ancestor of All Life

Inner-City Homeless and the Genetics of Resistance

For the next 3 billion years, the only living things on Earth were one-celled organisms, the archaea and prokaryotes. Like the first RNA and the multicelled organisms that later evolved, these one-celled animals went through their own distinct evolutionary process. Cell biologists and biochemists have analyzed cell fossils and have been able to reconstruct the most likely evolutionary paths taken over the last 3 billion years. Independent cells first lived by photosynthesis, making energy from sunlight. In the next major evolutionary hurdle, cells adapted from an atmosphere with virtually no oxygen to one that gradually became rich in oxygen over the 500 million years after cells first appeared. Cells that survived this increase in oxygen and began

FIGURE 16-4
Typical Bacterial Cell

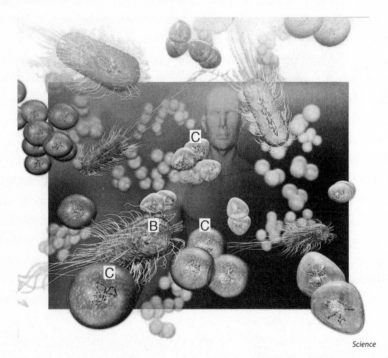

Science

Two common forms of bacteria (prokaryotes) are seen here, the round cocci (*C*) and rod-shaped bacilli (*B*). The bacilli are covered with hair-like projections called pili, which play a role in increasing the organisms' virulence. In the foreground, we can see that the largest coccus has its thread-like genetic material (DNA) winding through the organism, not enclosed in a nuclear membrane. Thus it does not have a true nucleus like the cells of more complex animals. Just above the largest bacillus in the foreground, two new cocci are forming by binary fission (or splitting) from a single coccus. These organisms range from about 2 to 10 micrometers in diameter (cocci) or length (bacilli). A human can see about 500 to 1000 micrometers with the naked eye (without magnification). Long chains of cocci can also be seen.

to use it to their advantage—a primitive form of respiration—multiplied to cover the early Earth.

These early cells evolved into bacterial species from which all other living things, including plant and animal life, evolved. Bacteria are one-celled creatures and have primitive forms of all the organelles described earlier in this chapter, but they do not possess a well-defined nucleus, thus the label as a prokaryote. It would take thousands of average-size bacteria to cover the period at the end of this sentence. Figure 16-4 shows typical bacterial cells. Some are rod-shaped and are called bacilli. Several species of these have rigid hair-like projections called pili covering their exterior membrane, and the pili play a major role in how dangerous certain of these bacteria may be. Other species, the cocci, are round and may live as single organisms or in long chains. Many species have a tail-like structure called a flagellum that contains contractile actin and microtubules and can whip about to propel the organism along.

In favorable conditions, some species of bacteria are able to reproduce every fifteen minutes. Early ancestors of present-day bacteria entered into complex relationships with algae (primitive plants) and formed mound structures, several yards high and across. These simple ecosystems, the fossils of which are called stromatolites, covered the shallow coastal waters of the Earth. The same types of formations are common today and are found in the shallow waters of the Gulf of California and along the Great Barrier Reef off the northeast coast of Australia.

Human and animal bodies are teeming with thousands of different species of bacteria (commonly called microbes), and the resident populations of these organisms make up the "microbiome." We have evolved through the millennia with our own microbial communities. Humans have been found to house at least 1,000 different species of bacteria in their gastrointestinal tract alone, and any given individual is expected to have about 160 bacterial species. In addition, the reproductive tract, skin, mouth, and nose all harbor numerous species of bacteria, numbering in the millions. The millions of bacterial genes present on and in our bodies (our microbiome) vastly outnumber the approximately 20,000 genes that make us human (see Chapter 19). But fortunately, due to natural selection, a relatively small number of bacterial species cause disease. Most are actually our friends and are harmless or may be *vital* to the existence of life, including our own. Various species are essential for digesting our food, enhancing our immune systems, and keeping disease-causing organisms from invading our tissues.

For example:

- Bacteria in cows' stomachs are essential for digesting the cellulose in their food. Similarly, termites require a full complement of specific bacterial types in their intestinal tracts to digest the wood they consume.

- Organic waste substances in sewage and other organic materials are degraded by bacteria and transformed into material that plants depend on for growth.

- Certain species of bacteria cause cheese and other dairy products to be formed by metabolizing constituents of milk.

- Some bacteria add needed nitrogen to soil.

- The coliform bacteria living in our intestines are essential for a comfortable balance of digestion. If we go to a foreign country and ingest different bacterial types, we can suffer an imbalance in our digestive process.

Of course, the best-known bacterial types are those that cause disease. These pathogens can infect virtually every region of the human body. For example:

- Meningococcal bacteria infect the spinal and brain membranes.

- Diphtheria-causing bacteria have a unique affinity for the throat.

- Tubercle bacteria invade the lungs.

Bacteria, like all other organisms, continue to evolve according to the principle of natural selection. For example, the deadly lung disease tuberculosis, caused by the bacterium *Mycobacterium tuberculosis*, has reemerged as an unfortunate chapter in medical history due to the ongoing process of natural selection in this bacterial strain. By the middle of the twentieth century, it appeared that tuberculosis was essentially wiped out by treating infected individuals orally with a combination of two drugs, rifampin and isoniazid. Any person can be infected by tuberculosis, but those who have a normal immune system usually do not get the disease because they are able to kill the bacteria by natural defense mechanisms or to keep the bacteria walled off in small scars in the lung, preventing the tiny organisms from growing and destroying lung tissue. If a person's immune system cannot keep the bacteria isolated and in check, they multiply in great numbers, thereby causing overwhelming infection and destruction of the lung tissue. The drugs mentioned stop this disease process only when the treatment program is followed faithfully and for many months.

In the mid-1980s, there was a dramatic increase in the number of homeless people, those addicted to intravenous drugs, and sufferers of AIDS—many of whom have dangerously depressed immune systems. Because *Mycobacterium* thrives in human hosts with such depressed immune systems, tuberculosis made a comeback in inner-city populations. The lifestyle of those groups of highly susceptible individuals whose numbers increased beginning in the mid-1980s and continuing to today are not conducive to a complete course of drug therapy to clear the infection. In other words, those most likely to be infected do not take their medicine properly or effectively, thereby providing a window of opportunity for the bacteria.

Some mycobacterial organisms possess genes that rendered the drugs ineffective and have thereby allowed natural selection of those organisms to take place, resulting in new strains of antibiotic-resistant *Mycobacterium* that cannot be killed by any known drug. As a result, thousands of new cases of tuberculosis are being diagnosed every day as the newly adapted and infectious bacteria are spread by coughing and close contact, particularly in large cities worldwide. New drugs are currently being developed, but many people will die before the genetics of resistance is understood and effective therapeutic strategies take hold once more. According to the World Health Organization, in 2011, approximately 500,000 new cases of tuberculosis were diagnosed in children, with millions of additional cases in adults worldwide, particularly in Asian and African countries.

Certain bacteria can survive independent of other organisms, while others are parasites. *Viruses*, on the other hand, are never fully independent and are not cells. Each is a genetic system dependent on cells to survive. In other words, viruses need to grow inside animal, plant, or bacterial cells. They're much smaller and simpler than bacteria, consisting of only nucleic acid (RNA or DNA), a coat of protein, and in some cases fatty and carbohydrate material.

FIGURE 16-5
Typical Viruses

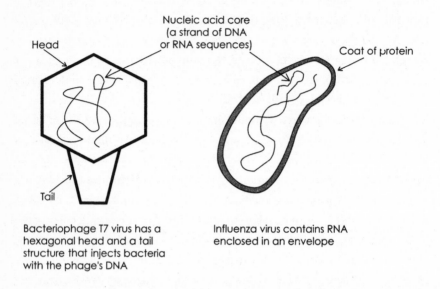

Bacteriophage T7 virus has a
hexagonal head and a tail
structure that injects bacteria
with the phage's DNA

Influenza virus contains RNA
enclosed in an envelope

Some viruses called bacteriophages (or bacteria eaters) insert a spike-like device into bacteria to use the larger organism's enzymes and metabolic processes to reproduce. Though certain species of viruses and bacteria cause disease in humans and virtually all other species, obviously there is no harm "intended." They are simply living in the particular environment in which they evolved and adapted, reproducing and dividing like any other organism.

The Primordial Cell Evolves into All of Life

Evidence in Slime Molds and Sponges

After one-celled organisms first appeared on Earth, millions of additional chemical and molecular changes developed under the pressure of natural selection over the next hundreds of millions of years, including the appearance of the DNA molecule (see Chapter 18), which gradually took over the self-replicating duties of RNA. Single-celled protozoans and other one-cell organisms obviously were enormously successful in evolutionary terms, and still are. But looking at humans and at all the successful *multicell creatures,* or metazoans, we see it became advantageous for an organism to consist of more than one cell. Thus about 600 to 800 million years ago, multicellular organisms first appeared. Through examination of the fossil record and today's creatures, the particular course of development from one-celled to multicelled life is fairly clear. We can see strong evidence of that evolutionary pathway in the organization of the "lower phyla," beginning with simple multicelled animals that evolved from the single-celled protozoans. For example, the organism we know as the slime mold exists at certain times of its life cycle as a single cell that can move around by means of a flagellum (long hair-like structure) just like many protozoa. At other times, many of these individuals gather together, their cell membranes coalesce, and a large mass containing thousands of nuclei moves about functioning as a single organism ingesting food. To replicate, this mass of protoplasm forms a platform with a stock that produces spores, much like a plant. This fascinating creature has features of both plant and animal cells, and scientists are not sure just how to classify it. There is little doubt that the cells that constitute the slime mold represent a very early life form that recapitulates much of what we understand about the first self-replicating cell on Earth.

As another example of early evolution from single to multicelled organisms, *Volvox* proliferates in freshwater habitats about the globe. *Volvox* is simply a collection of individual single-celled protozoans living together in an adherent colony. Obviously, this configuration provided the *Volvox* with a survival advantage and marks an early evolutionary swing toward the successful multicelled organism.

This pathway can be seen easily in the next most complex life forms that we know as the poriferans, or sponges. There are hundreds of different kinds of sponges, but each one consists of *three simple cell types* that are organized to carry out specific functions and that live together as a single organism: one cell type that moves water around an open body cavity, one that secretes a silica-based tissue for support, and one that digests small food particles drifting by. If the three cell types that make up a sponge are broken apart, they are capable of recognizing "self" and will stick together again in their own primitive way, but only to each other. These cells are not yet organized into tissues, but each of these cells has a nucleus and the organelles described earlier.

We need not travel very far up the evolutionary ladder before we find cells that make up tissues—that is, groups of interacting cells of similar structure and function, like skin, muscle, and nerve. These can be found in the jelly fishes, anemones, and starfish. As simple as they are, these creatures can feel and many are quite mobile. Starfish even have primitive cells organized to detect light at the tips of their arms.

The next steps up in evolutionary complexity are the flatworms, the roundworms, and then the segmented worms (Figure 16-6). The arthropods (insects, crustaceans, etc.) followed several hundred million years later, and the amphibians, reptiles, and mammals were not far behind. The evolutionary pressures of natural selection were slowly creating thousands of new, more complex species. Yet each cell of every organism functions on the same principles of energy production and replication as those of the protozoans, slime mold, *Volvox*, and sponges.

Knowledge of cell function and cell evolution constitutes a major aspect of our evolution. It is also the basis for understanding reproduction and the origin of all life in the primordial soup.

FIGURE 16-6
Flatworm and Segmented (Earth) Worm

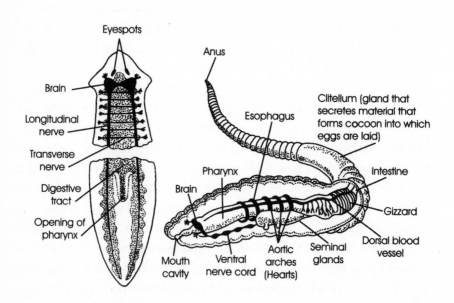

Mental and Physical Ascent of Humans from a Single Cell Is Astounding

Macrophage Outposts and Human Cells That Crawl

Over a seventy-year life span, the human brain holds about *100 trillion* bits of information. By comparison, the entire *Encyclopaedia Britannica* contains only about 200 million bits. The superior power of the human brain is a result of natural selection, in the exact same manner as any other traits that confer a survival advantage. As discussed in Chapter 14, it is clear that our early ancestors who wandered the African plains were already surviving more by guile than brute force or speed.

We are equally complicated physiologically. Of the 60 trillion cells making up each of our bodies, not only do we count among them the cells that are organized into tissues like nerves and muscle that in turn constitute organs such as the brain and heart but we also have millions of individual cells that we depend on for our survival. These single cells are found in the hemolymph (blood) of grasshoppers and circulating in the blood and on the lung surfaces and in liver tissue of dogs, cats, rabbits, giraffes, horses, and humans. They are the cells that provide essential defense mechanisms to protect us all from infectious viruses and bacteria and invasion by parasites. The similarity of our own cells (and those of grasshoppers) to protists that evolved millions of years ago is striking. An excellent example is the single-celled protozoan known as the amoeba compared to a cell that moves through our blood and around our body cavities with the sole purpose of removing particulates and organisms that do not belong. Amoebae are single-celled miniature "blobs" of cytoplasm with all of the organelles described earlier. They can be free-living, feeding on bacteria or fungal spores, or they can be parasitic like the species *Entamoeba histolytica* that causes severe gastrointestinal disease in people unfortunate enough to ingest the cells that may be living on lettuce or other vegetables. The cell that we and just about every other multicelled species has evolved with is called the "macrophage" (*macro* meaning "big" and *phage* meaning "eater"). If the cells are found in the lung, they are called lung macrophages, in the liver they are called liver macrophages, and so on. Their major function is to protect us from invading organisms like bacteria and parasites that they can kill with an armamentarium of enzymes and reactive oxygen molecules. In the lungs, they have the special adaptation of removing inhaled particles from the air spaces (called alveoli). Thus as we travel around the earth, we (and all the other species) inhale dust particles, bacteria, and pollen grains. The alveolar macrophages have multiple receptors on the cell surface that recognize the foreign particles; the cells crawl (like an amoeba) along the cells of the airspace surface, consume the inhaled dust particles and carry them off the breathing surfaces of the lungs, up the trachea, and ultimately out of the body where the cells and the ingested debris can be swallowed or spit out. Every one of our millions of alveoli normally has about one to three macrophages waiting for inhaled particles. Cigarette smokers have hundreds of these cells in every air space to deal with the thousands of particulates inhaled with every breath of smoke. The two macrophages seen in a human air space in Figure 16-7 strike different poses because the cell in the foreground has detected the presence of a pollen grain or bacterium that was inhaled and was caught in the act of migrating toward that particle when it was photographed with an electron microscope by the author. The other macrophage is covered by a ruffled membrane and will respond within minutes if attracted by an inhaled particle or microorganism.

New macrophages are constantly being produced in the bone marrow. The immature cells move from the bone marrow into the blood flow, and if they exit the blood as it passes through the lung, these cells will mature as lung macrophages. If they reach the liver, they become liver macrophages, with all the special features these cells exhibit. Other types of macrophages and phagocytes (cells that eat or pick up particulates) course through our blood vessels; gather dead blood cells for demolition; and continuously defend us against infection by bacteria, fungi, and parasites. We have an immune system that depends on these amoeba-like cells as well as white blood cells called lymphocytes that make antibodies to bacteria and viruses, all designed to keep us in good health.

This diverse array of single cells is part of the class of cells that constitute our body's defense mechanisms. As seen by the examples, they engulf and destroy bacteria, viruses, and other foreign substances as well as worn-out and abnormal cells. They originate in the bone marrow and then venture out to assume their appointed duties at numerous outposts throughout the body,

FIGURE 16-7

Two Alveolar Macrophages in the Human Lung

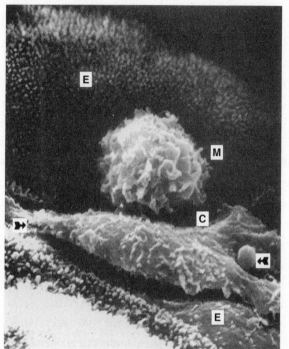

A. R. Brody

Macrophages are cells that patrol our airspace surfaces and pick up particles that do not belong there. These macrophages (*M*) are seen with an electron microscope in a human lung, and the elongated one in the foreground (*C*) is about to surround the small round particle (*arrow*) (a coccus or pollen grain), pick it up in a process called phagocytosis (cell eating), break it down in internal enzyme sacks, and then carry it up the airways and ultimately out of the lung to be swallowed or spit out. Every air-breathing animal (rats, cats, elephants, giraffes, horses, etc.) has these cells in the lungs and in other organs for protection. The macrophages are sitting on epithelial cells (*E*) that line our airspace surfaces, and the macrophage in the background remains round, ruffled, and ready to attack foreign particles that may be inhaled.

as just described. Although they exhibit many characteristics similar to bacteria and the cells of the mites living in the pores of your forehead, and they look and behave similar to the free-living one-celled amoeba and protozoans, they are a product of *human* genes. In other words, our own DNA is programmed to develop these "lively animalcules." They don't join us later, like bacteria, mites, the amoeba, and protozoan parasites.

As stated earlier, the clear unity among life's diversity is seen in cells. Amazingly, if cells from any tissue or organ in the human body (skin, kidney, heart, lung, etc.) are removed and placed in a dish with the correct nutrients, each cell will immediately begin to act like a single-celled organism. They will crawl, take up the required nutrients and many will reproduce by cell division. As each cell moves by contraction of the actin cytoskeleton, it searches the space ahead of it and can pull back or bind to other like cells, ultimately forming rudimentary tissue in the culture dish.

With 100 trillion bits of information in our brains and with a makeup and function that produces one-celled entities that police our bodies as part of an elaborate defense system, we are incredibly complex beings, intellectually and physically. Even if one fully subscribes to the 65-million-year evolutionary process (from ancestral primates into *Homo sapiens*) described in Chapter 14, it might still be difficult to accept the idea that if you go back farther in time we have actually evolved from a one-celled organism. How could that simple cell be the universal ancestor? Yet, in Part Five we saw that acceptance of great time spans was necessary to understand Darwin's theory. Within the same huge time span the process of natural selection resulted in the evolution of one-celled organisms into multicelled life. Emotion and myth about human origins must give way to the facts we have in our grasp today.

Was the development of complex organisms like humans inevitable or did it occur by chance and historical accident? This question has been debated for decades and has led to many refinements of evolutionary theory, such as Stephen J. Gould's theory of punctuated equilibrium and the emerging theories about self-organization and the "science" of complexity. Entire books are devoted to the principles of self-organization and complexity, such as *At Home in the Universe,* by Stuart Kauffman, in which he contends that there are general laws and principles of order that apply to complex biological systems and, in combination with natural selection, make the appearance and evolution of life inevitable. The question of inevitability versus chance is a philosophical question that might never be answered conclusively. However, at the beginning of the twenty-first century we have come to understand the cell and its workings. We see that scientists have been able to determine the biological processes that occurred over those billions of years and pushed the original one-celled organisms to evolve into complex plants, animals, and humans.

Humankind has always been fascinated by its place and role in the universe and by its beginnings. The birth of the universe. The beginning of life. The first humans on Earth. Einstein once said that Copernicus taught us to be modest by showing that the Earth and humankind were not the center of the universe. Darwin said we should be proud for having ascended to the pinnacle of all forms of life, the only species that is self-conscious, contemplates its fate, and understands history. Now, scientists have been able to reconstruct the mix of elements that formed the first organic molecules, and trace evolutionary history from the first cells to 60 trillion cells working together. Whether or not this evolutionary path was inevitable or accidental, we are now in the twenty-first century, and biochemists and biologists have explained to us the essence of our very existence.

Chapter Seventeen
BEADS ON A WIRE

For Weismann, the germ-cells are immortal. . . . Each generation hands to the next one the immortal stream unmodified by the experience of the body. . . . The body is transient-temporary. Its chief "purpose" is not its individual life, so much as its power to support and carry to the next point the all-important reproductive material. . . . [But] the germ-plasm must sometimes change—otherwise there could be no evolution. . . . Is there some internal, initial or driving impulse that has led to the process of evolution? . . . We can only reply the assumption of an internal force puts the problem beyond the field of scientific explanation.
—Thomas Hunt Morgan, *Heredity and Sex* (1913)

The city of Brno sits at the confluence of the Svratka and Svitava Rivers in what is now the Czech Republic, formerly the southern Moravia region of Austria. Founded in 1243, it survived the ravages of the Hussites, the Bohemians, the Swedes, the Silesian War, and Napoleon, and became the site of a now-famous series of experiments that resulted in the first steps toward understanding how biological traits and characteristics are transmitted from parents to their offspring.

Mendel Formulates the Basic Principles of Genetics

Uncut Pages

The folklore concept that heredity is passed on through our blood is reflected in a number of common terms, such as "bloodlines," "half blood," "mixed blood," "bad blood," "blue blood," "royal blood," and "blood relative." Once again, we can look to Aristotle for the origin of this mistaken belief, yet it was accepted through the nineteenth century by most biologists, including Darwin. (Ironically, mature red blood cells are the only cells in our bodies that do not contain genes because those cells have no nuclei.) Gregor Mendel (1822–1884) played a pivotal role in dispelling the old beliefs about inherited traits and establishing the study of heredity as a biological science. Born in Heinzendorf, Austria, Mendel was the son of impoverished peasant parents and developed an interest in agriculture in his youth on the small family farm and orchard. He entered the Augustinian monastery in Brno in 1843 and was ordained a priest four years later. The Augustinians paid for and supported his continued interest in the natural sciences, sending him to the University of Vienna where he studied physics, chemistry, mathematics, zoology, and botany. He returned to Brno in 1854 and began to teach natural science at the high school, where several colleagues were engaged in scientific pursuits and later founded the Natural Science Society. Mendel joined that group and was able to expand on his particular interest in botany through the many science books available at the school and monastery libraries.

Beginning in 1856, Mendel began experiments crossing varieties of pea plants in the small garden at the monastery to observe the statistical results of combining seven characteristics of the plants, including seed color, seed shape, tallness versus dwarfishness, and four other factors. For example, he wanted to see whether the offspring of a tall pea plant that was crossed with the pollen of a short pea plant would be of tall, short, or medium height. Over the next several years of experiments in the garden and analysis of literally thousands of pea plants, he determined the numerical ratios of various characteristics of the plants that resulted from the crosses. From those ratios he established five principles that apply equally to all living things and endure to this day:

1. Each physical characteristic of a living organism (from one-celled protozoa to human beings) is the product of a specific "hereditary factor," which Mendel envisioned as some type of particle (now known as a gene).

2. These hereditary factors exist in living things in *pairs*. For example, the mother's pair might consist of a factor for green eyes and a factor for hazel eyes, and the father's pair, green and blue.

3. With respect to each such characteristic, only one of the two factors (for example, for green eyes) that exist in the mother and one of the two (say, for blue eyes) in the father are passed to each of the parent's offspring.

4. There is an equal probability that either one of the mother's factors and either one of the father's factors will be inherited by each of their offspring. So, following the example, the result could be any one of these combinations of hereditary factors for eye color in each child:

From Mother/Father
> Green/Green
> Green/Blue
> Hazel/Green
> Hazel/Blue

The offspring's eyes and all of its other physical characteristics are never a *mixture* of factors.

5. Some factors are dominant and some recessive. Therefore, if the factor inducing green eyes were dominant over the blue or hazel factors and if the blue eyes factor were dominant over the hazel factor, 75 percent of the offspring of the two hypothetical parents would have green eyes (that is, the first three examples in number 4), and 25 percent of them would have blue eyes (the last example). The only way for the child to have hazel eyes would be if both parents provided the recessive "factor" for hazel eye color.

In 1865 (six years after publication of Darwin's *On the Origin of Species*), Mendel presented two papers (titled "Experiments with Plant Hybrids") to the Natural Science Society in Brno explaining the results of his pea plant research. Yet, no one in this learned body recognized the application that Mendel's findings could have to understanding inheritance or evolution. Mendel even pointed out to them that no scientist had previously made it possible "to determine the number of different forms under which the off-spring of hybrids appear, or . . . to ascertain their statistical relations." To the other members of the society, his work was mathematics, which they thought had little application to botany or natural philosophy. His paper was published in 1866 along with the other proceedings of the society for the previous year, and the entire volume was routinely distributed to academic libraries in London, Paris, Vienna, Berlin, Rome, and other major cities throughout Europe, as well as in the United States. Mendel's paper went largely unread and unnoticed, except by the church authorities in Moravia, who admonished him for what appeared to be Darwinian views and reminded him of the church's stance on evolution. Mendel's last eighteen years were uneventful from a scientific point of view. He returned to his monastic work, was elected abbot of his monastery in 1868, and was honored and respected for other work in the community. Meanwhile, his revolutionary paper lay dormant on obscure library shelves in those cities throughout the world.

Though Mendel collected all of Darwin's articles and books before and after *On the Origin of Species*, there's no evidence that the two men ever corresponded with each other. Even if they had, it's difficult to say what might have come out of such contact, but it probably would not have altered the course of the history of heredity as a biological science, particularly because specific evidence of Mendel's hereditary "factors" had to await the work in cell research discussed in Chapter 16, which led to the birth of genetics itself in the twentieth century.

Other chapters in the story begun by Darwin with *On the Origin of Species* were slowly and quietly being written in the latter part of the nineteenth century by Walther Flemming and August Weismann. As mentioned in the previous chapter, Flemming's highly accurate description of cell division in his 1882 book, *Cell-Substance, Nucleus, and Cell-Division*, showed that cell reproduction involves the transfer of chromosomes from the parent to daughter cells through mitosis. Weismann (1834–1914), a German biologist, was familiar with Flemming's findings and recognized that they did not answer his questions about the division of the unique cells known as egg and sperm. With Mendel's experiments unknown to Weismann and the rest of the world, Weismann wrote that the sperm and egg cells of animals must contain "something essential for the species, something which must be

FIGURE 17-1
Stages of Meiosis

Meiosis is the Unique Process of Cell Division
for Egg Cells and Sperm Cells

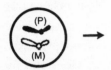 →  → →

1. Paternal (black) and maternal (white) chromosomes contract to form in cell nucleus

2. Prophase–Replication of genetic material occurs, followed by pairing of chromosomes with the same kinds of genes, which then bond and exchange genetic material. This exchange is called recombination. (See Figure 17-2)

3. Metaphase– Paired chromosomes line up at "equator"

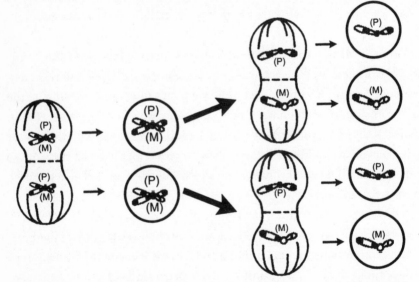

4. Anaphase I– Chromosomes separate in first cell division

5. Telophase I–Two new cells separate and nuclear envelope reforms

6. Anaphase II– Chromosomes separate in second cell division

7. Telophase II– Resulting in four cells, each with half the number of chromosomes (23) as in mitosis, and each with genetic material from both parents as a result of recombination

carefully preserved and passed on from one generation to another." This was his "germ plasm" theory, first published in *The Germ-Plasm: A Theory of Heredity*, in 1886, which concluded that all living things contain a special hereditary substance. He also recognized that if *all* the hereditary substances from the two parents became mixed together in the fertilized egg, there would be a geometric increase of those substances, which eventually would be impossible for the cells to store. Thus, as part of his theory, Weismann predicted that the material of heredity is carried by Flemming's chromosomes and that there is a form of cell division that includes nuclear division (now called "meiosis") resulting in each daughter nucleus receiving only *half* of the paired chromosomes from each of the parent cells' nucleus.

As the nineteenth century came to a close, a number of biologists were conducting research to test Weismann's theory and gain insight into the behavior of the sperm and egg cells during this process. Some of that research involved experiments much like Mendel's of thirty-five years earlier. It was at this point—on the threshold of the twentieth century that was to come forth with more new knowledge than in all of previous scientific history combined—that Mendel's classic paper was rediscovered, retrieved from its scholarly grave in March 1900 by Hugo de Vries of the Netherlands through the publication of his two papers on the hybridization of plants. In one of the papers, de Vries noted that Mendel's study "is so rarely quoted that I myself did not become acquainted with it until I had concluded most of my experiments, and had independently deduced the above propositions." When other scientists went to the library shelves to read their copy of the 1865 *Proceedings of the Brno Society for the Study of Natural Science*, they found that the pages of Mendel's paper had not been cut. That is, no one had separated the edges after the volumes were sent to the various libraries. *The paper had never been read.* For thirty-four years it had not seen the light of day, literally and figuratively, but the words on those pages were soon to be transformed from obscure musings into the basic laws of heredity.

The Science of Genetics Is Born

Mapping Chromosomes

With Mendel's ideas introduced into the mainstream of research on heredity, the significance of Flemming's mitosis and Weismann's meiosis became apparent. In a 1902 paper in the *Biological Bulletin*, later refined in an article titled "The Chromosomes in Heredity," the American geneticist Walter S. Sutton (1877–1916) provided the first conclusive evidence that *chromosomes carry the units of inheritance and that they occur in distinct pairs*, as Weismann had predicted. The evidence was obtained through Sutton's experiments on sperm formation in grasshoppers in which he observed that each chromosome in the cell nucleus becomes paired with a physically similar chromosome. As shown in Figure 17-1, the paired chromosomes then contract, duplicate, and later separate, with each going to a different sperm or egg cell. The physical pairing of the paternal and maternal copy of each chromosome is unique to meiosis. In comparison, in mitosis (as seen in Figure 16-3), the paternal and maternal chromosomes that have the same genes (called homologous chromosomes because they are the same) line up with each other, go through a replication phase, and then are separated into the new daughter cells. In meiosis, the maternal and paternal chromosomes replicate, bond to go through recombination (exchange of genetic material), and then separate. In mitosis, each cell (lung, liver, skin, etc) produces two new cells with a full complement of paired chromosomes, whereas the process of meiosis involves *two separate cell divisions* resulting in *four* new cells (the egg and sperm) each with half the number of chromosomes as all the other cells in our body and in the bodies of every creature that procreates. The simple illustration (Figure 17-1) shows only two chromosomes in the egg or sperm nucleus; however, the nucleus of every human egg and sperm cell contains twenty-three chromosomes, while the nucleus of all our other cells has twenty-three *pairs* of chromosomes.

Meiosis differs from mitosis in another most important way. The resulting "contents" of the new egg and sperm cells, the chromosomes with their genes arranged in a specific order, are solely responsible for the hereditary makeup of any offspring resulting from the union of the two cells. While mitosis is the process through which all cells, of all living things, are maintained at normal numbers or multiply to compensate in case of injury, meiosis involves the fundamental processes of genetics and evolution. Thus Sutton's experiments on meiosis converted Mendel's abstract mathematics and Weismann's predictions into physical reality by identifying the "collections" of these particles or materials (later called genes) on chromosomes as the hereditary factors.

Meiosis and the present discussion concern the division of egg and sperm cells. This is not about fertilization of the egg or reproduction of the entire organism (such as a human). However, once meiosis was understood, it became clear that the *essence of heredity is the duplication of genes, the carriers of genetic information.* Beginning at the moment the egg is fertilized, and the individual chromosomes can be in pairs once again, the process of mitosis (as discussed in Chapter 16) results in this one fertilized egg cell ultimately dividing into trillions of cells. "Directed" by the genes, these cells differentiate into the hundreds of different cell types that make up a functioning organism. In the case of humans, the cells have specific functions like liver and kidney cells, nerve cells, and skin and lung cells. In a worm, the cell types are fewer, but the concept is the same. As we will explore in the next chapter, it would still be decades after the work of Sutton and other geneticists before scientists were able to determine the specific chemical composition of genes.

Beginning in 1903 and continuing for decades, the following scientists expanded on the discoveries of Darwin, Mendel, Flemming, Weismann, and Sutton, and refined our understanding of the principles at work when offspring inherit their parents' genetic makeup:

- **Herman Nilsson-Ehle (1873–1949):** Beginning in 1900 and continuing to his retirement in 1939, this Swedish geneticist conducted research with wheat strains and other plants, refining and confirming Mendel's five principles of inheritance. Over the course of his career, Nilsson-Ehle opened up new fields of research in genes and chromosomes and advanced the knowledge of mutations.

- **Edward M. East (1879–1938):** His pioneering work in plant genetics and botany beginning in 1900 included experiments with tobacco plants, in which he accurately concluded that spontaneous mutations in the genes themselves accounted for certain changes throughout generations of such plants without any change in environmental conditions. These random mutations can be essential in the process of natural selection because when such a mutation confers an advantageous trait, it is selected for by an increased chance of that organism surviving. That mutated trait is then passed on to the offspring.

- **Thomas Hunt Morgan (1866–1945):** Through his experiments with the fruit fly *Drosophila* (which produce a new generation in two weeks), Morgan, a geneticist and zoologist, made the monumental discovery that chromosomes are *not permanent structures*. In 1909, it was Morgan who adopted the word *gene* (first used that year by the Danish botanist and geneticist Wilhelm Ludvig Johannsen) to refer to one of Mendel's "hereditary factors." Over the next several years, in collaboration with three of his graduate students at Columbia, Morgan not only confirmed Sutton's theory that each chromosome carried a collection of genes "strung out like beads on a wire" but discovered that the position of each of those beads could be mapped and identified within precise regions of the chromosomes (see Figures 18-3 and 19-1).

More important, Morgan and his group were first to prove that during the stage when chromosomes become paired and then contract, they may *exchange* genetic material between chromosomes of maternal and paternal origin, as noted in the prophase stage of meiosis (Figure 17-2). This is called crossing over. When the "recombined" chromosomes separate into two newly formed cells, those new cells contain a unique array of genes in their chromosomes as a result of that exchange.

Figure 17-1 illustrates this exchange of genetic material between chromosomes during meiosis. Figure 17-2 shows this concept in more detail, illustrating the strands of that material now known to be DNA that make up the chromosomes. Recombination of genetic material also occurs when single-celled bacteria or simple fungi reproduce by cellular division. However, the term *recombination* should refer to the exchange of DNA sequences (genes) between maternal and paternal chromosomes during the normal process of meiosis that occurs whenever egg and sperm cells are components of the reproductive process. The critical significance of recombination is that the

FIGURE 17-2

Genetic Recombination during Meiosis

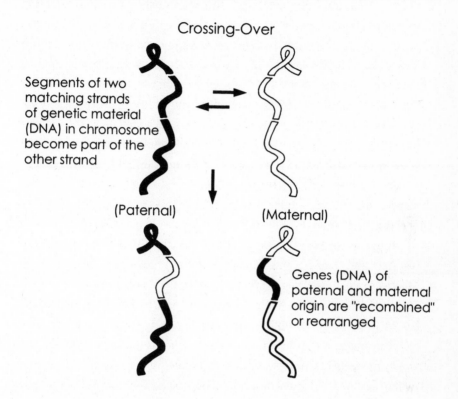

Crossing-Over

Segments of two matching strands of genetic material (DNA) in chromosome become part of the other strand

(Paternal)

(Maternal)

Genes (DNA) of paternal and maternal origin are "recombined" or rearranged

egg and sperm cells are, of course, the only cells that determine the complete genetic makeup of the offspring. Morgan and his colleagues proved that the process of variation, which circumstantially explains evolution, is *not* due to significant mutations occurring in every new generation, but to recombination of the "beads on a wire"—the genes. Thus Morgan established a firm relationship between Darwin and Mendel and discovered that Mendel's factors have a physical basis in chromosome structure. He also laid the foundation for the theory of linear arrangement of genes in chromosomes. In 1915, Morgan and his assistants published these findings in *The Mechanism of Mendelian Heredity*. In 1933, Morgan was awarded the Nobel Prize for this spectacularly prescient work in the rapidly expanding field of genetics.

Other extraordinary scientists moved this field rapidly to where we are today:

- **R. A. Fisher (1890–1962), J. B. S. Haldane (1892–1964), and Sewall Wright (1889–1988):** In the 1920s, these three geneticists, well versed in mathematics, independently but simultaneously calculated that the small variations from chromosome recombination as well as the spontaneous mutations deduced by Edward East could mathematically account for the vast changes in living organisms through the periods of time inferred by fossil evidence and required for evolution by natural selection. Six decades after the Brno Society for the Study of Natural Science had sorely missed the importance of Mendel's statistics, these three individuals founded the field of population genetics and gave natural selection a mathematical basis and explanation. Ronald Fisher's book in particular, *The Genetical Theory of Natural Selection*, published in 1930, showed that the slow but constant change in genes and chromosomes explains Darwinian evolution.

At age seven Sewall Wright wrote a booklet titled "The Wonders of Nature," and later became fascinated by genetics when he read about Mendel in the *Encyclopaedia Britannica*. His first paper was published in 1912 (when he was a graduate student, mentored by Edward East) and his last in 1988. By combining his interest in genetics with his formidable talent in math, which he taught himself, Wright performed pioneering work in mathematical population genetics and evolutionary theory, including his treatise on those subjects, *Evolution and the Genetics of Populations* (published in four volumes, 1968, 1969, 1977, and 1978). As the sole survivor of those who established genetics in the early twentieth century, Sewall Wright's death in 1988 marked the end of an era in which the fundamental "unit of heredity" was established.

- **Barbara McClintock (1902–1992):** After receiving her PhD in botany from Cornell in 1927, this American geneticist taught botany and genetics for several years and then took a research position at the Cold Spring Harbor Laboratory in New York where she performed a series of experiments on the colors of corn kernels that provided new and definitive information about recombination, the reality and characteristics of linkage groups of genes, and the relationship among particular genes. She remained at Cold Spring Harbor for fifty years and won the Nobel

Prize in 1983 for her pioneering work done in the 1940s and 1950s on genetic function and organization.

Today, we are in the era of the Human Genome Project, discussed in Chapter 19, in which scientists have now identified the precise sequences and locations of the close to 20,000 genes that it takes to be human and their relationship to one another on our twenty-three pairs of chromosomes. The entire genomes of a number of plants and insects as well as those of the rat, mouse, chimpanzee, and gorilla also have been sequenced (see further discussion to come).

Once the gene had been conclusively identified as the essential heredity factor postulated by Mendel, the focus turned from hereditary principles and concepts to the physical and chemical components of this mysterious and extraordinary bit of matter in the cell's nucleus. By 1932, James Chadwick had discovered the neutron, the third subatomic element in the *atom's* nucleus, and ten years later physicists learned to harness the immense power of the atom. Curiously, however, it would take many more years before we fully understood the structure and inner workings of the cell's nucleus, even though it is much larger, because it consists of *combinations* of atoms in the form of the crucial and highly complex organic molecules that control all life. Yet it was that very combination and complexity that caused the mystery to persist. Amazingly, in the early years of the twentieth century, at almost precisely the same time that Ernest Rutherford predicted that the forces of the atom's nucleus might never be fully understood, Thomas Hunt Morgan said of genes in the cell's nucleus that "an internal force puts the problem beyond the field of scientific explanation." A generation later, in the seventh and most recent great scientific discovery—which will be explored in the following pages—this "internal force" was identified and its structure and function became understood, thus closing the circle of knowledge on the three interrelated and basis aspects of all life: *its origin, reproduction, and evolution.*

PART SEVEN

The Structure of the DNA Molecule

In Focus

This is the last of the seven greatest scientific discoveries in history—the configuration of deoxyribonucleic acid, commonly known as the DNA molecule. In turning from Part Six to Part Seven we are advancing from cell structure and principles of genetics to the physical components and inner workings of certain molecules in the cell's nucleus, including DNA. Because all aspects of living things are ultimately determined at this molecular level, it was necessary to gain an understanding of the molecule of life to understand life itself. In Part Seven, we will see how we came to that understanding. We will see the structure and function of the organic molecule that is responsible for the existence of all life and is the mechanism by which evolution takes place. We'll recount the events that led two men to discover that structure in February 1953, less than a century after publication of *On the Origin of Species*.

Genetics brought together diverse disciplines, such as biochemistry, biophysics, and microbiology. Today, this combination is called molecular genetics or molecular biology—the study of the molecular structure of genes and the way genes control cells. In this all-encompassing field, we have now amassed an extraordinary amount of knowledge about the process that creates the molecules of life and the process by which those molecules direct the formation and creation of the rest of the organism. In fact, we have gained so much knowledge of how we are "made" that scientists have been able to "map" the entire human genetic code and apply this knowledge in attempting to cure a multitude of diseases. We will examine this amazing revolution and the enormous potential it holds for our lives.

Chapter Eighteen
THE BACKBONE OF LIFE

No new scientific laws are required to explain living organisms' complexity and organization. . . . The essence of life . . . is the ability of living things to extract energy from their environment and to use that energy both to build up their own complex structures and to copy themselves.
—John Gribbin, *In Search of the Double Helix* (1985)

The book of life is very rich; a typical chromosomal DNA molecule in a human being is composed of about five billion pairs of nucleotides. The genetic instructions of all other life on Earth are written in the same language with the same code book. . . . This shared genetic language is one line of evidence that all the organisms on Earth are descended from a single ancestor, a single instance of the origin of life some four billion years ago.
—Carl Sagan, *The Dragons of Eden* (1977)

Exactly what are genes and what are they made of? In search of the answer to this question, scientists first explored the chemical and molecular composition of genetic material, then its most basic atomic structure. As Albert Lehninger stated in *Principles of Biochemistry*, "Living things are composed of lifeless molecules [that] conform to all the physical and chemical laws that describe the behavior of inanimate matter." Yet the molecules that provide the essence of life have a unique and extraordinarily complex structure. Understanding the complexity of this structure, including the particular functions of the molecules and how they sustain life through reproduction, became a fascinating and elusive goal of scientists all the way through the twentieth century.

Identification of Nucleic Acids and the Right Molecule Launch Molecular Genetics

Suggestion of a Secret Code

Decades before geneticists such as Thomas Hunt Morgan and Barbara McClintock expanded our knowledge of the organization of genes and chromosomes, a number of chemists were exploring the makeup of the cell's nucleus. They did not realize that their work would ultimately provide key information about heredity. In 1869, the Swiss biochemist Friedrich Mieschner (1844–1895) first proposed that all cells' nuclei probably have a specific chemistry. In succeeding years he discovered various substances in the nucleus that he separated into *protein and acid molecules*; hence the term *nucleic acids.*

A Russian-born chemist, Phoebus A. T. Levene (1869–1940), was also a pioneer in the study of nucleic acids. After receiving his MD degree from the St. Petersburg Imperial Medical Academy in 1891, Levene fled anti-Semitism in Russia and settled in New York City where he studied chemistry at Columbia University and then engaged in chemical research from 1905 to 1939 at the Rockefeller Institute for Chemical Research. In 1909 he correctly identified ribose as the sugar in one of the two types of nucleic acids, ribonucleic acid (RNA), and he identified certain components in the other nucleic acid, deoxyribonucleic acid (DNA). But Levene and many of his colleagues were convinced that the complex and plentiful *protein molecules* (and not DNA) stored all the genetic information in the chromosomes. Unfortunately, researchers labored under this false assumption for decades as they pursued the source of some type of gene-copying mechanism in the proteins. Levene's theory of the purpose of DNA—that it merely holds protein molecules together—turned out to be incorrect.

The work that led to correcting that false assumption was begun in 1928 by Frederick Griffith (1881–1941), an English bacteriologist. His research on pneumonia-causing bacteria (pneumococci) led to his discovery that a certain unknown substance from the cells of one strain of *dead* pneumococci was able to enter a different and living strain and cause the live strain to pass the dead strain's hereditary characteristics to the live strain's offspring. Another bacteriologist, Oswald T. Avery (1877–1955), together with his colleagues, became aware of the significance of Griffith's work and spent ten years trying to identify the agent that was the essence of genetic transformation in the bacteria. Finally, in 1944 Avery and his collaborators published the results of their lengthy research, which clearly showed that it was the DNA, not protein or RNA, that allowed the transport of hereditary information. This work launched the science of molecular genetics.

Subsequent research in the United States by the Austrian-born biochemist Erwin Chargaff (1905–2002) determined the proportions of the four principal molecules present in DNA: adenine (A), cytosine (C), guanine (G), and thymine (T). The term in chemistry for these substances is *base*, meaning something that reacts with acids to form salts and has certain other chemical characteristics. Chargaff determined the exact proportionate amounts of the DNA bases within each molecule: *guanine equals cytosine* and *adenine equals thymine*. Therefore, the amount of guanine and adenine *combined* equals that of cytosine and thymine *combined*. These are known as the "Chargaff ratios" and were to become a key element in the discovery of the structure of the DNA molecule.

A series of experiments in the 1940s and early 1950s by the American biologist Alfred D. Hershey (1908–1997), dealing with DNA of bacteriophages—viruses that infect bacteria—further confirmed the Avery group's conclusion that DNA, not protein, is the genetic material. In 1969, Hershey won the Nobel Prize for this work. Thus the work by the early geneticists, as well as Avery, Chargaff, and Hershey, suggested that there was some sort of genetic information or code configured in DNA and passed on to offspring of all living things.

As mentioned earlier, nucleic acids come in two types: RNA and DNA. The bases are the same in both molecules, except that uracil replaces thymine in RNA. The chemical composition of DNA can be pictured as shown in Figure 18-1.

As we will soon see, the *bases* of DNA and RNA are of critical importance in understanding what a gene is.

FIGURE 18-1

Backbone and Bases of DNA

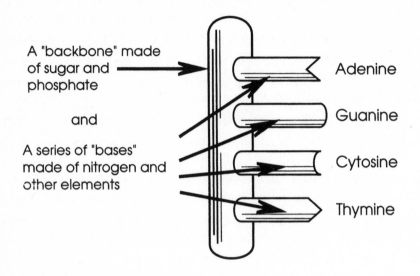

Genetics and Chemistry Combine with Physics to Bring Us to the Threshold of Discovery

Quantum Physics Revisited

The foundation of the next phase of this story dates back to the time of Röntgen's discovery of x-rays in 1896 and the work of Thomson, Becquerel, Rutherford, Bohr, Einstein, Planck, and others. As discussed in Part Two, these scientists discovered the present model of the positively charged atomic nucleus (protons and neutrons) with negatively charged electrons orbiting the nucleus within distinct shells. Aggregates of atoms form into molecules by sharing electrons in the outermost shell or orbit of each atom—the outermost shell requires the additional electrons, resulting in the chemical bond between elements. Bonds between hydrogen atoms are slightly different from other chemical bonds in that they do not actually share electrons, but are brought together by the electrical attraction of the atoms. The same principles of bonding that apply to simple molecules like H_2O (water) also apply to the much more complex organic molecules we are dealing with in biology, which contain billions of atoms.

Those basic principles of chemistry and physics eventually combined with biology to give us the answer to the question asked earlier: *What is a gene?* In 1935, the German-born physicist Max Delbrück (1906–1981) co-authored a now-famous paper with two biologists that addressed the question of why short-wave radiation causes mutations in genetic material. Delbrück explained that the answer is found in the branch of physics called quantum theory, because mutation in living organisms involves molecules in the cell's nucleus being pushed over an energy barrier (measured in quanta) from one stable configuration to a different stable configuration. In other words, he demonstrated that the physics of the atom itself determines the most basic biological processes. Influenced by Delbrück's paper, Erwin Schrödinger (1887–1961), the famous Austrian theoretical physicist, wrote the book *What Is Life?* in 1944, in which he further illustrated how quantum physics explains the stability of the molecules in the genetic structure. In 1944, the results of Oswald Avery's work on DNA had just been published, so Schrödinger's book did not state that genetic information was contained within DNA. However, it had a great impact on several other physicists working in biology who later recognized the connection between DNA and Schrödinger's calculations.

Other physicists began to recognize the importance of Delbrück's and Schrödinger's groundbreaking work. In addition, the biochemists Linus Pauling (1901–1994) and J. D. Bernal (1901–1971) were making significant discoveries about the structure of proteins. Influenced by these developments, after World War II ended in 1945 a number of physicists turned away from fundamental physics to the exciting and burgeoning field of biology, where they saw a major role for physics in solving biological issues. They brought with them the thought processes characteristic of their field—that is, the concept that all material bodies (such as organic molecules like DNA and RNA, made up of the organic bases) are arrangements of large numbers of atoms, and the behavior and property of these bodies can be explained by determining how atoms are joined together to form molecules.

For a physicist to make this transition to biology was like a doctor becoming an architectural engineer—the fields seemed to have nothing in common. It was a bold decision. The vast majority of physicists and biologists didn't recognize the potential benefits in developing such an alliance. However, this handful of researchers in both fields began to engage in molecular biology, and the term *biophysics* was coined. In its infant stages, biophysics was merely a technique to apply the knowledge about atoms and molecules to biological issues, but it soon became an important part of molecular genetics.

One person influenced by Schrödinger and his work was Maurice H. F. Wilkins (1916–2004), a New Zealand–born British biophysicist who used x-ray diffraction studies to examine the molecular structure of DNA. X-ray diffraction (also called x-ray crystallography) is a phenomenon in which the regularly arranged atoms of matter (particularly a crystal structure) diffract or scatter the x-rays, which can then be recorded on a photographic plate, much like Wilhelm Röntgen's original discovery of x-rays in 1895. The pattern that shows up on the photograph provides important clues to the configuration of the atoms in the material, which would otherwise not be detectable because until very recently atoms were too small to be seen with any type of microscope. Wilkins first gained knowledge of x-ray diffraction through his use of the mass spectrograph in his two years of work on the Manhattan Project during World War II. After his return to England, he and a research student named Rosalind Franklin (1920–1958) used x-ray diffraction to take pictures of fibers composed of pure DNA. These pictures became a critical link in the chain of events that ultimately led to the discovery of the structure of the DNA molecule.

By the late 1940s, molecular biologists knew that DNA was the molecule of life, but they did not know its exact structure and, therefore, how it carried out its functions or how it duplicated itself. However, the time was ripe for the work in genetics to converge with the new knowledge of the chemistry of the cell's nucleus. This convergence would reveal DNA's secrets. By applying the newly found information from biophysics to these other bodies of knowledge, two men fiercely pursued the structure of DNA in the hope they could entice it to whisper its secret into their ears before anyone else discovered it. This fateful and serendipitous pursuit resulted in one of science's, and indeed history's, most magnificent moments.

Crick and Watson Meet and Begin Their Collaboration

Basing a Career on the Gossip Test

Francis Harry Compton Crick (1916–2004) was born in Northampton, England, the older son of a middle-class couple. His father, Harry Crick, and his uncle ran a boot and shoe factory founded by their father in Northampton, a town dominated by the manufacturing of footwear. At an early age, he showed an interest in scientific subjects, so much so that his parents bought him a set of Arthur Mee's *Children's Encyclopaedia* when he was eight because they were unable to answer his constant questions about the universe. "I read it all avidly," Crick later wrote. "It was the science that appealed to me most. What was the universe like? What were atoms? How did things grow?" Yet, as he marveled over the fascinating facts in the encyclopedia, he was struck by one great fear: "By the time I grew up . . . everything would have been discovered." When he confided in his mother about this fear, neither knew how prophetic her reassurance would be. "There will be plenty left for you to find out," she told her son. By his own description, Francis Crick was not exceptionally precocious or outstanding in school, but his curiosity persisted. This desire to know the answers to the great questions he'd posed as a child, together with an unusual reasoning ability and affinity for logical thinking, were to chart his destiny.

After graduating from University College in London in 1937 with a degree in physics, he engaged in postgraduate research in hydrodynamics. This provided him with a great appreciation of the scientific method and with the idea that a variety of scientific disciplines could be as successful as physics had been in the early part of the twentieth century. During World War II, Crick worked in the Admiralty Research Laboratory, designing magnetic and acoustic mines. After the war, at age thirty, he was faced with the decision of what to do with the rest of his life. As he was to write later, he concluded that "what you are really interested in is what you gossip about." He realized he was not gossiping about physics but about recent advances in biology. Using this unscientific methodology to determine his future and considering his desire to engage in fundamental research, Crick narrowed his interests to "the borderline between the living and the non-living." He was referring to the fact that no one had yet fully explained how lifeless atoms could create life or how organisms pass on their characteristics. Convinced that the gossip test had led him to his true calling, he became one of the handful of adventurous postwar physicists to cross over to biology.

In 1946 Crick took a job with the Strangeways Laboratory in Cambridge, England, to study the physical properties of the cytoplasm of cells. After about two years there, he came to the realization that "one needed to discover the molecular structure of genes" to understand life and that "the most useful thing a gene could do would be to direct the synthesis of a protein, probably by means of an RNA intermediate." At that time, Crick learned that a research unit dedicated to studying the structure of proteins using x-ray diffraction was about to be established at the Cavendish Laboratory of Cambridge University, the site of many earlier great discoveries in physics. In 1949, at the age of thirty-three, Francis Crick transferred to this lab to become a graduate student again and learn about the three-dimensional crystalline structure of proteins, with the hope that this great laboratory's fame in physics could be extended into biology.

Meanwhile, James D. Watson (1928–) had obtained his doctorate from Indiana University at the tender age of twenty-two and accepted a U.S. government fellowship to work under the biochemist Herman Kalckar in Copenhagen. By his own admission he possessed the grandiose and vain notion that he would discover the secret of life. "My interest in DNA had grown out of a desire, first picked up while a senior in college, to learn what the gene was. . . . It was certainly better to imagine myself becoming famous than maturing into a stifled academic who had never risked a thought." At the forefront of the research on x-ray diffraction of DNA were biophysicist Maurice Wilkins and his assistant, Rosalind Franklin, at the newly created biophysics department at King's College in London. After attending a lecture by Wilkins in the spring of 1951, Watson "began worrying about where I could learn how to solve X-ray diffraction pictures." Without approval from the Fellowship Office in Washington, DC, he left his fellowship appointment to take a job working on x-ray crystallography at Cavendish, where Crick had been since 1949.

Cavendish was under the direction of physicist Sir Lawrence Bragg

(1890–1971), who had won the Nobel Prize in 1915 for his own x-ray diffraction work. This included the formulation of Bragg's law, an equation that *correlates the wavelengths of x-rays and the distance between "planes" of atoms*. At Cavendish, Watson was supposed to work on determining the complex structure of the protein myoglobin, a molecule in muscle. However, like Wilkins and Crick, James Watson hoped that x-ray diffraction would help reveal the particular configuration among atoms in DNA.

According to Robert Shapiro, in *The Human Blueprint*, Watson arrived in Cambridge "as a socially awkward, untidily dressed, nervous, ambitious, and intense young American PhD." In the fall of 1951, Francis Crick met this brilliant and brazen young researcher. They soon learned they shared a passion for the elusive macromolecule DNA, both being convinced it was the secret of life, thus beginning the union that was soon to lead to the most recent of the seven greatest scientific discoveries in history. Crick later wrote:

> Jim and I hit it off immediately, partly because our interests were astonishingly similar and partly, I suspect, because a certain youthful arrogance, a ruthlessness, and an impatience with sloppy thinking came naturally to both of us. Jim was distinctly more outspoken than I was, but our thought processes were fairly similar. . . . I knew a fair amount about proteins and X-ray diffraction. Jim knew . . . about the experimental work on phages (bacterial viruses) and . . . bacterial genetics.

Ironically, neither Crick nor Watson were officially working on DNA at Cavendish. "I was trying to write a thesis on the X-ray diffraction of polypeptides and proteins," Crick explained, "while Jim had ostensibly come to Cambridge to . . . crystallize myoglobin."

The Search Goes under Cover, Then Becomes a Race to Beat the Americans

The Pauling Factor

At first, Crick told Watson that solving the x-ray diffraction patterns of the DNA fibers should be left up to Wilkins and Franklin. However, they soon became impatient with the other pair's slow progress, which was largely the result of Wilkins's and Franklin's dislike for each other. Thus Crick and Watson undertook analysis of the diffraction patterns on their own. However, Crick did not get along well with his boss, Sir Lawrence Bragg. Also, Wilkins's and Franklin's director at King's College convinced Bragg that Crick and Watson were duplicating the King's College work. So in the fall of 1951, soon after Crick and Watson began their joint work, Bragg ordered them to cease their unofficial quest for the secret of life and return full-time to the mundane research they were supposed to be pursuing at Cavendish— the structure of the myoglobin protein. Without rank and without any concrete results with which to dissuade Bragg, their only course of action was to pursue their obsession in secret. As Watson later explained in his book *The Double Helix*, they could not even chance a casual question to Maurice Wilkins because that "would provoke the suspicion that we were at it again." At first, they limited their work to discrete conversations at lunch. Soon, however, as Watson explained, he began to spend "the dark and chilly days to learn more theoretical chemistry or to leaf through journals, hoping that possibly there existed a forgotten clue to DNA."

But in January 1952, even this remnant of the secret DNA project was threatened with total dissolution when the Fellowship Office in Washington, DC, revoked future Nobel laureate Watson's stipend and award because he'd left Copenhagen for Cavendish and demanded his immediate return to the United States. Not to be distracted from his stubborn pursuit, Watson defiantly wrote back that he found Cambridge intellectually exciting and that he had no immediate plans to return to the States.

Spurring the pair on with more intensity were the stories coming out of the California Institute of Technology, where the famous biochemist and eventual three-time Nobel Prize winner Linus Pauling appeared to be close to discovering the structure of DNA. Linus's son, Peter Pauling, was a research student at Cavendish while Watson and Crick were there and exacerbated their fears with frequent reports about his father's progress on DNA. But late one night in June 1952, Watson developed a negative of an x-ray he'd taken of DNA material. "The moment I held the still-wet negative against the light box," said Watson, "I knew we had it. *The tell-tale helical markings were unmistakable.*" Crick, the physicist and the expert in x-ray diffraction of macromolecules spotted the crucial twist of the crystal structure in "less than ten seconds," thus confirming Watson's earlier suspicion that DNA had a helical, not linear, configuration.

Nevertheless, by the end of 1952 they "remained stuck at the same place we were twelve months before" because they still didn't have proof of how many strands were in the helix or how to explain its structure from a chemical and molecular standpoint. That is, they couldn't explain how or why the components of DNA (the sugar-phosphate backbone and the four organic bases—adenine, guanine, cytosine, and thymine) all fit together. Thus when Peter Pauling came into the lab one day in December 1952 and announced that his father had determined the structure of DNA, their hearts sank. They feared the race had been lost and the Nobel had slipped away. But as weeks went by without any excited announcements or publications coming from Pauling and Caltech, Watson and Crick thought it unlikely that Pauling could have discovered the actual DNA structure, particularly because he didn't have access to the Wilkins–Franklin x-ray work at King's College.

In the end, Peter showed his father's manuscript to Crick and Watson, and to their immense relief it showed a *three-chain* helix with a sugar-phosphate backbone in the *center*. They had independently toyed with the idea of a three-chain molecule over a year earlier and were convinced it was not accurate. "Everything I knew," Watson recalled, "about nucleic-acid chemistry indicated that phosphate groups never contained bound hydrogen atoms. . . . Yet somehow Linus, unquestionably the world's most astute chemist, had come to the opposite conclusion. . . . We were still in the game."

However, they knew that when Pauling became aware of his mistake, he would pursue this work at an even faster pace to find the right structure. So, for the first time since Bragg had banned them from DNA, Crick and Watson openly pleaded their case to him—it was now imperative that they get back into the DNA business with both feet to beat out the scientists on the other side of the Atlantic. Crick and Watson estimated they had six weeks before Pauling would realize the errors he'd made in the three-chain theory.

Wilkins's X-Rays and Quantum Mechanics Lead to the Final Answer

No Experiments and a Wood and Metal Model

Professor Bragg finally saw the immense potential in their work and with at least a temporary reprieve, the first order of business was to find out the latest results of Wilkins and Franklin's work at King's. When Watson walked into Wilkins's office in January 1953, he saw to his amazement that Wilkins and Franklin had the clearest evidence of not only a helical structure but vital parameters that could enable Watson and Crick to fix the number of DNA chains in the helix at two. Though the x-rays were a year old, Wilkins and Franklin failed to recognize their significance and were unable to decipher the helical aspects that were so clear to Watson and Crick. As Watson explained, "The instant I saw the picture my mouth fell open and my pulse began to race."

Watson raced back to the Cavendish to begin building a two-chain model. "Important biological objects come in pairs," he observed. After hearing what Watson had learned at King's, Bragg began to cheer him on in the race against Pauling, while Crick continued to tread more lightly because he was not fully back in Bragg's good graces. The two proceeded up numerous blind alleys over the next several weeks until the American crystallographer Jerry Donohue convinced them of a particular quantum mechanical argument about hydrogen bonding and the pairing of adenine only with thymine, and guanine only with cytosine (based on the Chargaff ratios) that sent them in the right theoretical direction. In late February 1953, the day after they spoke with Donohue, Watson suddenly realized that "an adenine-thymine pair held together by two hydrogen bonds was identical in shape to a guanine-cytosine pair." This bonding and pairing solved the last remaining problem they'd faced because it accounted for the two types of base pairs being identical in shape so the double helix structure would be made up of uniform angles and rotation. This was the crucial revelation that enabled them to make the final mathematical calculations and begin constructing a wood and metal model.

Incredibly, during their eighteen-month collaboration (from fall 1951 to

March 1953) Crick and Watson never performed experiments to determine the structure of the DNA molecule. They performed limited research in areas related to their pursuit of the double helix such as chemical base pairing and taking x-ray photographs of the tobacco mosaic virus, but no experiments on the structure of DNA. They engaged only in theoretical discussions on the subject and built the models. Reflecting on the relatively short period of time actually devoted to discovering the structure, Crick stated:

> We had one intensive period of model building toward the end of 1951, but after that I myself was forbidden . . . to do anything further, as I was still a graduate student. For a week or so in the summer of 1952 I had experimented to see if I could find evidence for bases pairing in solution. . . . The final attack [in February and March, 1953] . . . only took a few weeks. Hardly more than a month or so after that our papers appeared in *Nature*.

The first of a series of four articles appeared in *Nature* on April 25, 1953, and described the structure of DNA and explained that it *controls the manufacture of proteins*, as suspected by Crick in 1948. The essential aspects of the double helix, as described in the articles, are shown in Figure 18-2—the two helical chains, running antiparallel, with the sugar-phosphate backbone on the outside and the bases (adenine, thymine, guanine, and cytosine) on the inside. Because of the angles at which the DNA's chemicals bind to each other, all DNA molecules consist of two parallel twisting strands, like the handrails of a spiral staircase—thus, the label that immediately became famous with the Crick-Watson discovery: *the double helix*.

Figure 18-2 is simply the acid shown in Figure 18-1, *plus its mirror image*. Note that adenine will only "fit" into thymine, and guanine and cytosine similarly match up because of their particular chemical attractions—that is, the hydrogen bonds. This is the clearest demonstration of the need for biology and physics to join forces as biophysics in order to understand the structure of DNA. The paired bases are "stacked," as shown. Billions of stacked bases configure themselves into disk-like forms that make up each chromosome of every cell of every living creature. In the interphase stage (discussed in Chapter 16) when cells are not dividing, the DNA material (called

FIGURE 18-2
Double Helix

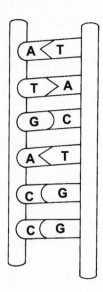

chromatin) floats around in the nucleus as a double helix, but without being organized into chromosomes. When cells receive a growth signal in interphase and begin to divide, the DNA condenses and coalesces into the structures that we recognize in the prophase as chromosomes (as shown in Figure 16-3).

Understanding DNA

DNA Makes RNA Makes Protein

As with the other great scientific discoveries, DNA can be clearly understood on a conceptual basis without the detailed knowledge of the molecule's physics, chemistry, or angles of diffraction that were so important to Crick and Watson. The points about the DNA molecule being a *double* helix and that it *duplicates* itself are always emphasized in discussions about DNA, while its primary role is often not adequately explained or recognized. To understand DNA, one has to understand the capability or function that is being duplicated. Exactly what does DNA do? *It controls the design, structure, and function of the entire organism by directing the manufacture of proteins, and it can be precisely duplicated so that each new cell that is formed is exactly like the original.* This function is equally applicable to a bacterium, protozoan, or a cell from a human lung or liver.

Proteins are made solely of amino acids (as distinguished from the nucleic acids that make up DNA and RNA). Amino acids are built around the four bonds of the carbon atom. That is, carbon has a valence of four, meaning it has four unpaired electrons in its outer shell, enabling it to form such bonds and making it the most important atom and chemical element in biology. Though there are only twenty varieties of amino acids, long repetitions of multiple sequences allow for tens of thousands of combinations of amino acids into a huge variety of proteins. In fact, there are about 50,000 different types of proteins in each of our bodies, and most of those types are unique to the human species. The same twenty amino acids in 50,000 different combinations are linked together in long chains folded up on themselves.

When amino acids join together to form proteins, they serve as the basic building blocks of life. Proteins are not simply a beneficial substance we obtain from meat and other foods. They are complex molecules that exhibit an extraordinary array of properties and functions as components of structural elements such as collagen (connective tissue which holds our organs together), hormones (which control division and behavior of cells), hemoglobin (which carries oxygen), and antibodies (which circulate in the blood to protect us from infections); as essential enzymes that trigger chemical reactions and changes throughout the body; and as catalysts in the DNA molecule itself. Remember that the genes of insects, mice, and elephants make essentially the same proteins by the same universal DNA copying mechanisms. In view of the complexity and pervasiveness of proteins in every cell, it's not surprising that Phoebus Levene and other biochemists were convinced for many years that genes were made of proteins rather than DNA.

Again looking at the big picture—the cells' physical operations outlined in Chapter 16—Figure 18-3 shows the relationship among the key components in the cell's nucleus.

Figure 18-3A shows DNA helices wound tightly during cell division into what we recognize as a chromosome. Red blood cells lose their nucleus shortly after being produced in the bone marrow and therefore have no

FIGURE 18-3

A. Drawing of an Electron Micrograph of Human Chromosome 12

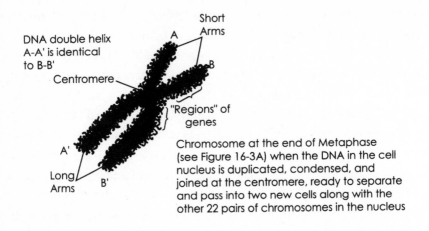

DNA double helix
A-A' is identical
to B-B'

Centromere

Short Arms

"Regions" of genes

Chromosome at the end of Metaphase (see Figure 16-3A) when the DNA in the cell nucleus is duplicated, condensed, and joined at the centromere, ready to separate and pass into two new cells along with the other 22 pairs of chromosomes in the nucleus

B. The Human Karyotype

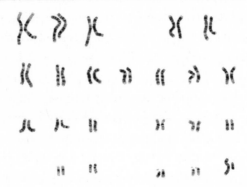

National Human Genome Research Institute

These twenty-three pairs of human chromosomes were photographed with a light microscope. One of the pair is from the female and the other from the male. During mitosis, each of these chromosomes will replicate as diagrammed in Figure 16-3A and photographed in Figure 16-3B. Chromosome pair 1 is in the upper left hand corner and the large X (female) chromosome and the smaller Y (male) chromosome are in the lower right corner. The chromosomes have been stained so that the light and dark bands that identify specific regions on the chromosomes have been revealed. The human X chromosome (*arrow*) is always larger than the Y, and the Y contains the *SRY* gene that controls the development of male characteristics.

chromosomes, and each human egg and sperm cell has twenty-three chromosomes—the haploid, or half the number of chromosomes. Apart from these exceptions in red cells and egg and sperm cells, the nucleus of every human cell contains forty-six chromosomes (the diploid number)—that is, twenty-two pairs of almost identical chromosomes, plus the XY chromosomes in males and the XX chromosomes in females (Figure 18-3B). Specific genes located on the relatively small Y chromosome are necessary for "maleness." Each human chromosome consists of a double helix with about 100 million nucleotides (that is, various sequences of bases A, C, G, and T) in each strand. Therefore, we can now more thoroughly answer the earlier question, What is a gene? *It is a region of DNA that controls the expression* of a specific hereditary characteristic like hair color, height, skin color, nose shape, and thousands of other traits that make us human, or in the case of a tick or elephant the inherited features that allow it to survive and prosper. The particular sequence of the organic bases (A, T, G, and C) that make up a given gene "codes for" the specific amino acids that form the corresponding proteins (see Figure 18-4). The DNA "sequence" (specific alignment of the bases) also controls the sequence of complementary RNA strands that will move out of the nucleus and control protein synthesis in the cell cytoplasm. Thus a gene is a specific sequence of bases that codes for particular amino acids that will be strung together to form proteins required for life. The functional unit of a gene also includes regulatory proteins, so-called transcription factors that control components of the complex processing of the DNA sequence to produce a functional protein. Just how this process works has filled the pages of molecular biology journals over the past two decades to the point at which it is now possible for scientists to sequence a particular gene from a mouse or a human, transcribe the protein made by that gene, and then use the protein to great advantage. A good example is the transcription of the human gene that codes for insulin. This gene can be transferred to the genome of billions of *E. coli* bacteria that then transcribe the insulin protein in sufficient amounts to treat diabetics.

Each chromosome is a chain or strand of thousands of genes linked end to end. The human genome (each person's total genetic makeup) is composed of about 20,000 genes—that is, 20,000 specific series of three-letter sequences. Each gene has a specific site on a specific chromosome and must be at that site on an intact chromosome to code for the correct protein. There is no mixing and matching of where our genes can be arranged for them to function properly. Abnormal arrangement or sequences of our genes is the basis for many diseases, including essentially all cancers.

Each DNA strand is about 600,000 times longer than it is wide. If straightened out it would be about *one yard long*. When the cell, nucleus, and chromosomes divide, each strand then serves as a template for formation of a new matching strand in each of the new cells because of the structure and base pairing discovered by Crick and Watson. This explains the second major aspect of DNA—the one we usually associate with the double helix—its ability to replicate itself. In other words, as the DNA is duplicated within each cell that is undergoing cell division (which is occurring by the thousands every second in our bodies), its ability to control the cell's and body's functions by directing the manufacture of proteins is also duplicated.

This brings us back to the main function of DNA: making proteins. Because the essence of life and heredity, the genes, evolved so that they are protected in the cell's nucleus (as shown in Figure 16-1), working copies of the genes are made in the nucleus and in turn leave the nucleus through specialized pores in the nuclear membrane and direct the manufacture of proteins in other parts of the cell, the cytoplasm. Thus a type of gene "blueprint" is needed. That blueprint is made by the other nucleic acid, RNA, which is made of A, C, G, and uracil (U), instead of thymine (T). RNA polymerase is the specific enzyme that has the ability to split DNA down the middle of the "steps." In other words, it "unzips" the bases down the middle—at their hydrogen bonds—and turns the double helix into two single helices with "half steps" exposed, breaking those bonds between the two strands that link A to T and C to G.

As will be discussed in greater detail in Chapter 19, a sequence of three of these four base pairs, called a codon, is necessary to form a "word" that stands for a specific amino acid, as shown in Figure 18-4.

Because the amino acids must be joined side by side to form proteins, the sequences of these three-letter codons along the DNA strands dictate the proteins that are unique to each of us.

■ One or more specific three-letter sequences of the bases (each called a codon) result in the ordering of the twenty amino acids.

FIGURE 18-4

Genetic Codes Specifying Amino Acids

DNA triplet	RNA triplet	amino acid		DNA triplet	RNA triplet	amino acid
AAA	UUU	phenylalanine (1)		ACA	UGU	cysteine (10)
AAG	UUC			ACG	UGC	
AAT	UUA			ACC	UGG	tryptophan (11)
AAC	UUG					
GAA	CUU	leucine (2)		ATA	UAU	tyrosine (12)
GAG	CUC			ATG	UAC	
GAT	CUA					
GAC	CUG			GCA	CGU	
				GCG	CGC	
AGA	UCU			GCT	CGA	arginine (13)
AGG	UCC			GCC	CCG	
AGT	UCA	serine (3)		TCT	AGA	
AGC	UCG			TCC	AGG	
TCA	AGU					
TCG	AGC			GTA	CAU	histidine (14)
				GTG	CAC	
GGA	CCU					
GGG	CCC	proline (4)		GTR	CAA	glutamine (15)
GGT	CCA			GTC	CAG	
GGC	CCG					
TAA	AUU			TTA	AAU	asparagine (16)
TAG	AUC	isoleucine (5)		TTG	AAC	
TAT	ADA					
				T TT	AAA	lysine (17)
TAC	AUG	methionine (6)		TTC	AAG	
TGA	ACU			CCA	GGU	
TGG	ACC	threonine (7)		CCG	GGC	glycine (18)
TGT	ACA			CCT	GGA	
TGC	ACG			CCC	GGG	
CAA	GUU			CTA	GAU	aspartic acid (19)
CAG	GUC	valine (8)		CTG	GAC	
CAT	GUA					
CAC	GUG			CTT	GAA	glutamic acid (20)
				CTC	GAG	
CGA	GCU					
CGG	GCC	alanine (9)		ATT	UAA	(termination: end
CGT	GCA			ATC	UAG	of specification)
CGC	GCG			ACT	UGA	

■ Amino acids combine in specific order to form the 50,000 different types of proteins in the human body. *Each combination of codons that is ordering the accumulation of specific amino acids into proteins is a gene.*

■ All 20,000 human genes are configured in the forty-six human chromosomes that are located in every nucleus in every cell (except red blood cells). They become coiled into the recognizable chromosomal form (Figure 18-3) during cell division.

In forming these codes, the RNA polymerase moves along the DNA molecule, unzipping it and allowing RNA molecules that are loose in the nucleus (and have the same base sequences) to coalesce and match up along the now-exposed A, C, G, and T points of the original strands of DNA. In fact, the RNA forms an exact transcription of the DNA (except that uracil, instead of thymine, matches up with adenine). This copy is called messenger RNA, as shown in Figure 18-5.

When the RNA polymerase gets to the molecular "stop sign" that's at the end of every gene, it falls off along with the newly made messenger RNA, which proceeds out of the nucleus over to one of the many ribosomes in the cell. The ribosome reads the RNA's message (spelled out in the three-letter words) and, according to the particular sequence of bases in the codon, assembles a series of amino acids from the stockpile floating around loose in the cell. This action creates "from scratch" a particular protein "written out"

FIGURE 18-5
Formation of Messenger RNA from the DNA Helix

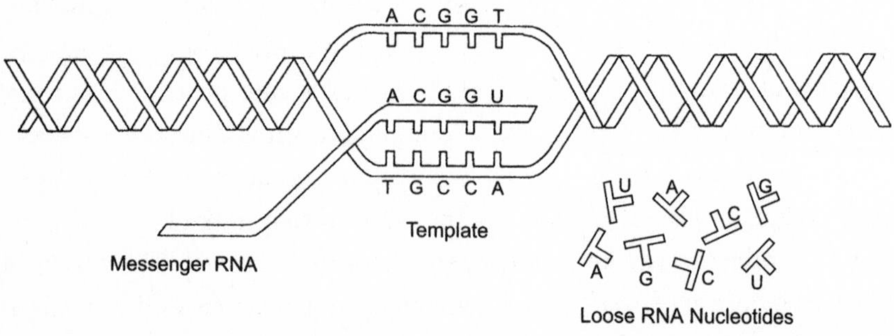

A C G G T

A C G G U

T G C C A

Template

Messenger RNA

Loose RNA Nucleotides

in language encoded originally by the sequence of three-letter bases in the DNA that remained in the cell's nucleus. Each of these new proteins reflects a small portion of the long DNA strands that contain all the three-letter codes for the thousands of different proteins. Accordingly, each cell in the body has different sets of genes "turned on," meaning that skin cells are synthesizing the proteins that keep our skin flexible and waterproof, yet the skin cells also contain the genes that code for digestive enzymes made only by liver cells. In turn, liver cells have active genes that code for enzymes that are pumped into our gastrointestinal tract to digest our food, while the genes in our liver cells that make skin proteins remain inactive, or "turned off." Molecular biologists like to think of our 20,000 some-odd genes as sort of a symphony orchestra where certain sets of genes are "playing" while others remain silent, depending on when and where the proteins they code for are needed.

Just as the RNA polymerase crept along the exposed DNA's G-C and A-T base-pairs to create the messenger RNA, the ribosome creeps along the messenger RNA to create a protein. Step-by-step, every vital protein formed in our bodies is made this way. At this very moment, thousands of ribosomes in *each* cell in your body are carrying out *millions* of reactions that are causing the amino acids listed in Figure 18-4 to bind into about 2,000 new protein molecules *every second*. As each protein leaves the ribosome and emerges from the cell, it has a particular folded and twisted shape, dictated by the chemical binding of the amino acids of which it is made. This shape and chemical makeup enables the 50,000 different kinds of proteins to perform their specific functions in the body.

We've now reached the climax of understanding the essence of heredity. *Because the nucleic acids (DNA and RNA) direct the manufacture of proteins, and the protein sequence is unique for each person, it is DNA that ultimately controls all hereditary characteristics.* Each three letter code instructs the assembly of a particular amino acid (Figure 18-4) into proteins in insects, mice, and humans. However, every living organism is different and every person unique because of the particular order or sequence of those three-letter codons originally dictated by the DNA of that individual human or beetle. In other words, the coding sequences that cause the formation of hair on a mouse are similar, but not identical, to the sequences that result in hair on a human's head. Likewise, the coding sequences that result in hair being formed on two humans' heads have greater similarity to each other than the mouse's sequences, but still are not identical.

This is the key to understanding hereditary and the function of DNA, and the reason molecular biologists refer to the phrase *DNA makes RNA makes protein* as the central dogma.

Crick credited Watson with the discovery that led to understanding the structure of the molecule—namely, Watson's pairing of adenine with thymine and guanine with cytosine. Watson, on the other hand, gave Crick much credit for recognizing the significance of the Chargaff ratios, concerning the relative proportions of the bases, and for recognizing the importance of pinpointing forces in the molecules by which like attracts like. That is, Crick identified the scheme for gene *replication* during cell division. Said James Watson, "Francis had the feeling that DNA replication involved specific attractive forces between the flat surfaces of the bases."

Thus in the first *Nature* article (about how DNA controls protein production), Crick and Watson made the ultimate scientific understatement: "It has not escaped our notice that the specific pairing we have postulated immediately suggests a possible copying mechanism for the genetic material." Five weeks after publication of that first article, Crick and Watson published a second paper in *Nature*, this time on the molecule's ability to duplicate genetic material.

Years later, Crick wrote that the ideas needed to grasp the structure of DNA

> are ridiculously easy, since they do not violate common sense. . . . I believe there is a good reason for the simplicity of the nucleic acids. They probably go back to the origin of life. . . . At that time mechanisms had to be fairly simple or life could not have been started. The double helix is indeed a remarkable molecule. Modern man is perhaps 50,000 years old . . . but DNA and RNA have been around for at least several billion years. . . . Yet we are the first creatures on Earth to become aware of its existence.

The Crick and Watson discovery was the culmination of eighty years of research by scores of scientists. During their eighteen-month association,

Crick and Watson made their way through thirty or forty discernible steps and missteps on the road to the ultimate solution, each derived from or dependent on an existing scientific fact or theory and each attributable to a predecessor or contemporary—people like Bragg, Chargaff, Pauling, Donohue, Wilkins, and Franklin.

James Watson's brash prediction at age twenty-two that he would discover the secret of life, Francis Crick's desire to understand "the borderline between the living and non-living," and his mother's prediction that "there will be plenty left for you to find out" all were fulfilled in 1953. In 1962, Crick and Watson, along with Maurice Wilkins, shared the Nobel Prize for their work.

Chapter Nineteen
THE HUMAN GENOME

As we experiment with the passages of our genetic text, we may find that certain selections . . . work better than others. . . . A couple in the future might want their child to have two copies of a gene associated with musical talent, but to lose one that conferred greater susceptibility to environmental cancer.

—Robert Shapiro, *The Human Blueprint* (1991)

The work of Crick and Watson led directly to the realization that it is possible to read and interpret the genetic plan of any organism, despite the fact the full text of humans has 3 *billion* characters that make up the 20,000 genes. That realization has now culminated in our ability to decipher the exact genetic makeup of human beings, which carries with it the promise of disease prevention and other improvements in the human condition.

Knowledge of the Structure Leads to Reading the Code

Another Moon Shot

As we saw in Chapter 18, a gene carries its information through a chemical code or blueprint provided by the sequence of nucleotides G-C and A-T that specifies the production of proteins. In 1954 (the year after the Crick and Watson discovery), George Gamow—the same person responsible for the "nuclear droplet" theory of the atom and the person who first called his expanding universe theory the Big Bang—made a significant contribution to genetics. He was the first person to propose the concept of a genetic code being "written" in sets of three of the four bases, called "triplets" of nucleo-tides, now known as the codons. As also seen in Chapter 18, the four bases of the nucleic acids that make up the DNA molecule form codes that detail which of the twenty amino acids that make up all proteins will be aligned in a particular order. Using two letters at a time (out of A, C, G, T) would allow for only sixteen "words," or combinations. But sixty-four words can be made if *three* of the four letters are used, which is more than enough to direct the construction of the twenty amino acids that make up proteins. Beginning with the first primordial cell, this simple construction function and lan-guage evolved as a result of the natural selection process described in Parts Five and Six.

Figure 18-4 lists the DNA code for all twenty amino acids, including the resulting complementary RNA sequences. In humans, the body makes twelve amino acids (called nonessential amino acids), while the remaining eight (essential amino acids) must be obtained through our diet. Again, *DNA makes RNA makes protein* applies here. With the exception of some viruses, *the DNA sequence of three-letter words in all living things is copied by RNA*, the genetic messenger (in a process called "transcription"), *which directs the synthesis or "manufacture" of proteins* (a process called "transla-tion"). The twenty amino acids and the three-letter code that makes each one are the same in all living organisms, but the particular proteins that are made of the amino acids differ from one species to the next. Most types of proteins contain about a hundred amino acids per protein molecule. There-fore, several types of amino acids occur repeatedly in the protein. That is the reason there are about 3 billion characters (A, C, G, T) that make up our 20,000 genes. Each cell "knows" its genetic code, thereby enabling a multi-celled organism to function, simply by carrying out *instructions* needed to *assemble* the "project." If one can imagine fifty instructions to assemble a piece of furniture or a working motor, assembling a living organism (by millions of instructions) certainly is quantitatively different but conceptu-ally the same.

After Crick and Watson discovered the DNA structure, there was an explosion of work in genetics labs worldwide in an effort to read the instruc-tions. In 1956, the British biochemist Vernon Ingram (1924–2006) discov-ered that sickle-cell anemia results from a mutation in the nucleotide sequence that codes for one single amino acid (glutamic acid) in the protein called hemoglobin, which carries oxygen in our red blood cells. That discovery demonstrated the critical link between the code sequence and the functions of the body. It is now known that where the three-letter nucleotide word should have been GAG (which is the DNA code for glutamic acid, as shown in Figure 18-4), it is misspelled as "GTG" (which is the DNA code for valine). This mutation in one gene, resulting from the incorrect code for one amino acid in the hemoglobin protein, changes the shape of that protein's surface, which in turn limits its ability to carry oxygen. The red blood cells that contain these abnormal hemoglobin molecules take on a deformed sickle shape. The resulting oxygen deficiency can lead to damaged organs throughout the body, vulnerability to a variety of infections, and death. Sickle-cell anemia is the most common genetic disease among black populations, affecting millions of people throughout the world—all because of a tiny one-letter misspelling in the genetic text. On the other hand, this sickle cell trait actually provides a survival advantage by reducing the likelihood of contracting malaria infection. The deformed, sickle-shaped red blood cells are less capable than normal red blood cells of being infected by the malarial parasites. Thus, the trait persists among many populations that are, or were historically, ravaged by malaria.

Mutations can also occur as a result of correctly coded genes lining up at the "wrong" location on a chromosome. As we saw in Chapter 16, during the interphase and in the prophase stage of cell division, long coils of DNA are wound diffusely throughout the nucleus of the cell and begin to condense into chromosomes. It is at this stage that the individual coding sequences, the genes, can first be recognized as discrete bands on the chromosomes. Each specific gene must line up on the same chromosome in the same position on that chromosome every time (Figure 18-3). However, it is during DNA replication and chromosome segregation that certain genetic errors can occur if some genes that are supposed to line up on one chromosome end up on another or in the wrong place on the right chromosome. As first recognized by geneticist Edward East, these chromosomal mutations, as well as mutations in the individual three-letter codes, can also lead to an advantage or disadvantage in a cell's or organism's likelihood to survive, thus providing much of the basis for the process of natural selection.

When research by the English biochemist Frederick Sanger (1918–2012) enabled scientists to begin sequencing RNA in the 1960s, it became theoretically possible to understand all of the vast amount of information in DNA, not just isolated examples. This led to a growing interest in "decoding" the molecule so we could actually know the relationship between each gene and each bodily characteristic, including gene-based diseases. In 1975, Walter Gilbert (1932–), a biochemist at Harvard, was the first to apply a particular chemical treatment to DNA to break it into fragments and to recognize the usefulness this might have in reading the script. The following year, Sanger and his assistants at the Cavendish Laboratory in Cambridge, England, spurred on by the reports of Gilbert's results, independently developed their own method for sequencing DNA (the "dideoxy method"). Up to 300 base sequences could be read in a single day. Their publication of the full sequence of a particular virus's DNA (called phi-X-174) in *Nature* on February 24, 1977, was the first DNA genome—that is, the first analysis of the full genetic text of any organism. Through the dideoxy method, Sanger made it theoretically possible to determine all of the text that governs the heredity of any living organism, including humans. Yet, because of the much greater complexity of humans (we receive 3 billion characters from each parent) compared to a virus, much work lay ahead to make this a reality. In 1981, Sanger, Gilbert, and Paul Berg, a pioneer in cloning, received the Nobel Prize for their independent work in sequencing DNA.

Despite the major accomplishments of Ingram, Sanger, Gilbert, and others, the effort to decipher the code of the human genome remained fragmented until 1989, thirty-six years after Crick and Watson displayed their wood and metal model of DNA. In that year, a coordinated national effort, similar to the Manhattan Project and space project, was begun. None other than James D. Watson addressed a gathering of scientists in Washington, DC, at a meeting called Genome I, which marked the consensus among molecular geneticists and others that it was time to launch a comprehensive effort to read the full text of the human genome. Such a text would contain the complete blueprint and evolutionary history of the human species. With an objective that was as clear and inspiring as landing on the moon, the Human Genome Project was born.

The Human Genome Project Is Launched

Landmarks and Maps

Watson had remained active in DNA research from 1953 to 1989, including dividing his time between Harvard and the Cold Spring Harbor Laboratory, New York, from 1968 to 1977. Francis Crick, on the other hand, ended up at the Salk Institute in San Diego to focus on his new passion (determined, most likely, by the "gossip test")—the study of the brain and the nature of consciousness. In 1977, Watson had begun working at Cold Spring Harbor full time as its chief administrator, greatly enhancing its academic reputation over the next decade (as well as raising funds enabling it to increase its annual budget from $600,000 to $28 million) and laying the foundation for this laboratory to become the headquarters for the Human Genome Project.

After the Genome I meeting in 1989, Congress appropriated $3 billion ($200 million a year for fifteen years) for the Human Genome Project and named James Watson as director of the newly created National Center for Human Genome Research, under the auspices of the U.S. Department of Health and Human Services. This project began on October 1, 1990. About ten years later, in February 2001, the preeminent scientific publications, *Science* and *Nature*, published the first sequences of the human genome. Scientists involved in this huge project expressed amazement at the low gene count humans possess—close to 20,000 rather than the originally expected 100,000 genes. The reason for this apparent discrepancy is that the genomes of most species, including humans, have long sequences of "noncoding," sometimes called junk, DNA. In fact, the vast majority (more than 95 percent) of the DNA in the nuclei of our cells is noncoding, and this large amount of DNA caused scientists to expect a much larger number of genes that code for the amino acids. The function of noncoding DNA is currently an area of extensive research. But it is the coding sequences that have raised so much hope for the advent of personalized medicine. In other words, once a given individual's genome has been sequenced, defects that produce a wide variety of diseases could be corrected. Approximately 80 percent of the genes sequenced in the Human Genome Project have been assigned a function, and many of them are dedicated to coding for proteins in our highly developed nervous system. It is striking to realize that the number of our genes is significantly less than twice that of worms and flies, indicating the complexity of these organisms as well and leading many scientists to use the homologous genes in many animal species to teach us the functions of our own.

Even before the advent of the completion of sequencing the human genome in 2001, a number of important discoveries were made concerning the functions of specific genes:

- **December 1989:** Scientists at MIT discover a gene that they believe is crucial in the development of human immune defenses, referred to as the gene *RAG-1*, which stands for "recombination activating gene." The discovery sheds new light on the complexities of the immune system, which is vital to every aspect of human health and development.

- **May 1991:** Doctors at Johns Hopkins Children's Center in Baltimore pinpoint the stage at which an error occurs in a mother's unfertilized egg, causing a child to be born with Down's syndrome, the most common genetic cause of mental retardation. By analyzing a specific marker, called a DNA polymorphism, they determine the origin of an extra chromosome and identify the point in time when the chromosome division malfunctions.

- **August 1991:** A collaborative research effort involving scientists from the Johns Hopkins School of Medicine, the Cancer Institute in Tokyo, and the University of Utah identifies the gene that initiates colon cancer. The gene is called *APC* (for adenomatous polyposis coli). This discovery will allow doctors to detect a colon tumor at the earliest possible stage.

- **March 1993:** Researchers announce that Huntington's disease results from unexplained "genetic stutters," expansions in the size of a particular gene on chromosome 4, which add extra strings of the amino acid glutamine to the protein that the gene normally encodes.

- **August 1993:** Researchers at Duke University Medical Center announce that people born with a variant of a gene called *APOe* (for apolipoprotein E) are more likely to develop Alzheimer's disease by age seventy than will people who carry other versions of the same gene.

- **June 1995:** A University of Toronto team announces that a gene on chromosome 14 is responsible for as many as 80 percent of familial cases of Alzheimer's disease.

- **August 1995:** Researchers at the University of Texas Health Science Center report that the *BRCA1* gene plays a major role in breast cancer.

- **December 1995:** British scientists announce the discovery of a second gene linked to breast cancer, *BRCA2*.

- **February 1996:** Scientists identify the gene that encodes a variety of cell surface proteins that travel to the brain and help regulate body weight, and speculate that obesity results from a defect or mutation in that receptor gene.

- **March 1996:** Researchers at the Oregon Health Sciences University report that healthy liver cells transplanted into diseased livers produce the missing enzyme FAH (for fumarylacetoacetate hydrolase). This gives new hope for gene therapy directed at the liver, which would reduce the need for liver transplants.

- **March 1996:** Researchers at five major medical centers announce they found a gene that heightens the risk of kidney disease and other lupus-related disorders. The defective version of this gene codes for a protein (called an Fc receptor) that is less efficient in its immune function than a normal version of the gene.

- **April 1996:** Molecular biologists at the Seattle VA Medical Center, the University of Washington, and Darwin Molecular Corporation announce they have found the human gene that causes the symptoms of aging. They predict that as we further understand how this gene works, we could have the ability to arrest the aging process and modify the gene's involvement in triggering heart disease, cancer, and osteoporosis.

Hundreds of discoveries like these followed, and scientific journals report these events in great detail. But the ultimate goal of knowing the sequence and function of each of our genes is to be able to conquer the multitude of diseases that have a genetic basis, like cancer and immune system and metabolic disorders. Francis S. Collins, the director of the National Institutes of Health, reports that scientists and physicians today are clearly closer to being able to do so than ever before. In the 2011 issue of *Science* magazine, he cites a striking example of how knowledge of the human genome has affected the life of a young boy. Nic Volker developed inflammatory bowel disease (a not uncommon condition) at about two years of age. But his disease did not respond to any conventional therapy. He had multiple surgeries and numerous treatment strategies but continued to worsen. Scientists at the Medical College of Wisconsin examined the sequences of the protein-coding regions of every one of Nic's genes. Based on what had been published through the Human Genome Project, the investigators found a mutation in a gene known as *XIAP* found on the X chromosome. This gene had previously been shown to play a role in a severe blood disease that could be cured with a bone marrow transplant. Nic underwent this transplant in 2010, soon began eating solid foods, and remains apparently healthy today. The thinking is that the stem cells in the transplanted bone marrow, all with a normal *XIAP* gene, repopulated the boy's gastrointestinal tract with new intestinal lining cells, thus providing normal function. Without firm knowledge of the sequence of this key gene, recognizing the gene's defect and the concept of how to replace it would not have been possible.

Numerous whole genome sequences have been completed on our closest relatives the chimpanzee (sequenced in 2005) and the gorilla (in 2012) as well as on several other nonhuman primates, a number of fish, reptiles and birds, multiple viruses and bacteria and insects, and dozens of plant species. All of these advances in understanding genes and how they function have allowed us to draw a significant conclusion: It is no longer necessary to theorize or to speculate on whether or not evolution has occurred and is operating today. The genome sequences prove the common ancestry of all life on earth. In this regard, multiple genes of model organisms like fruit flies, mice, and chimpanzees are homologous (similar in derivation, structure, and function) to ours and will eventually teach us the precise functions of the human genes.

Each of the discoveries of specific genes noted here relates to one or more specific sites on a particular chromosome, for example, the defective gene that causes Huntington's disease is located on chromosome 4, and Lou

Gehrig's disease (amyotrophic lateral sclerosis) is the result of a deficiency in a protein coded by a gene on chromosome 21. Down's syndrome, an all too common condition in every human population, results from an extra (or extra partial) copy of chromosome 21. Figure 19-1 is a map of all the regions of genes on human chromosomes 19 and 21, with the short arm on the top called "p" after the French word *petit* meaning small, and the long arm called "q" (the letter after *p*) on the bottom (see all of the human chromosomes in Figure 18-3). The numbers represent a universal system for mapping known gene sequences on specific sites on each chromosome. These sites can be seen when the chromosomes contract during mitosis and special stains are used to produce bands of light and dark DNA (Figure 18-3). Each

FIGURE 19-1

Genome Map of Chromosomes 19 and 21

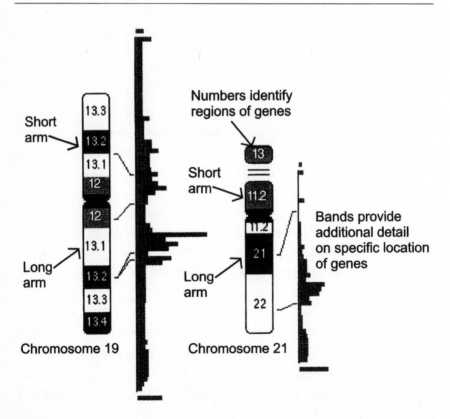

gene has a "locus," or specific position on a chromosome. For example, the locus of the gene known as *OCA1* is found at 11q1.4-q2.1. This gene, when not functioning properly, causes albinism (lack of skin pigment) and is located on chromosome 11, on the long arm (q), in the very short distance between band 1, subband 4, and band 2, subband 1. Thus the locus of every sequenced gene can be found across the human genome with this system. But more than that, the same method can be used to find the sequenced genes of a wide variety of species. In the field of research called synteny, scientists determine how the genes of other species match ours in sequence and function. As an example, it has been determined that chromosome 16 of the common house mouse (*Mus musculus*) contains multiple genes that are homologous with human genes found on chromosomes 3, 8, 12, and others. This "conserved synteny" between the mouse and human genes means that their origin and function are the same, demonstrating our common ancestry many millions of years ago. The sex chromosomes, X (female) and Y (male), have not been included in the numbering system, but these genes confer female and male characteristics throughout most mammals. While the pairing of XX makes us female and XY male, a more primitive mammalian ancestor like the platypus has ten sex chromosomes. The human Y chromosome is always smaller than the X (Figure 18-3B) and contains the *SRY* (sex determining region of the Y) gene. This gene, when expressed along with other male-specific genes, determines the male sex by coding for proteins that control the development of the testes. The X chromosome does not normally have the *SRY* gene; thus, an XX individual will produce hormones such as estrogen that lead to a female individual. There are several human diseases known as X-linked or Y-linked, which occur when certain genes on the X or Y chromosomes do not function properly and a condition is passed on to the offspring. Common examples of X-linked conditions are color blindness, muscular dystrophy, and hemophilia. There are comparatively few genes on the Y chromosome, but a form of male infertility has been demonstrated as a Y-linked defect.

Today the National Institutes of Health (NIH) includes the National Human Genome Research Institute (NHGRI), the newest of the twenty-five institutes that constitute the NIH. The NHGRI's website (www.genome.gov) is a fascinating place to wander about chromosomal maps and learn the astounding discoveries and daily advancements in personalized medicine and how scientists use model animal systems to understand normal human biology as well as disease processes.

The Project Spawns Hope, Fear, and Controversy

A Tale of Two Nuclei

Diseases caused by genetic mutations are a major source of disability, death, and human tragedy throughout the world. About 2 percent of all newborn babies suffer from genetic defects; 40 to 50 percent of all miscarriages are due to chromosome abnormalities. About 40 percent of all infant mortality is due to genetic disease. Around 30 percent of all children and 10 percent of adults are admitted to hospitals at some time in their lives because of genetic disorders. Cancers are the result of a series of genetic errors that cause loss of control of cell growth. The Human Genome Project was originally conceived and continues to be driven primarily by the hope of curing or reducing such diseases. The recent advances in genetic research demonstrate that this expectation was well founded and confirm the practical application of this work.

Yet the project has not been without opposition. Unlocking the secret in the cell's nucleus has given rise to a controversy remarkably reminiscent of the debate that ensued after physicists harnessed the power of the atom's nucleus. The genetic code is now understood to such an extent that redesigning the human genome and directing its instructions is feasible in the near future. Many people see great potential in applying this knowledge to cure diseases and improve the human condition, while others violently oppose such genetic engineering and gene therapy on ethical and scientific grounds. Indeed, in October 1993, Robert Stillman, a fertility specialist at George Washington University Medical Center, cloned human embryos, using methods common for breeding livestock and other animals. This was a laboratory experiment and was not done in connection with a pregnancy, but it did raise the possibility that identical twins could be "created" and born years apart. It also raised enormous ethical and legal issues.

Some people fear that the line will not be drawn at curing disease and that there will be an effort to genetically engineer "perfect humans" or better health, better physical appearances, or greater intelligence. Who would make the decisions? A wide array of activists, philosophers, scientists, and religious leaders has raised the specter of sliding past the cure of Huntington's disease and cancer into "fixing" color blindness and skin color, and perhaps even causing the extinction of the human race by tampering with the gene pool and altering evolution's natural course.

A number of years ago, a group made up of such diverse members as Pat Robertson, Jerry Falwell, Nobel laureate George Wald, twenty-one Roman Catholic bishops, and several Protestant and Jewish leaders issued a statement that science and society should not attempt to engineer specific traits into the human genome. They based their opposition primarily on the position that no person, group, or institution could ever "claim the right or authority to make decisions on behalf of the rest of the species alive today or for future generations." In a further link between the atom's nucleus and the cell's, this group's resolution included the statement that "Society should oppose human genetic engineering with the same courage and conviction as we now oppose the threat of nuclear extinction."

As indicated earlier, plants and animals commonly are cloned for our uses. Remember Dolly the sheep, the first mammal to be cloned in 1997? This was a scientific breakthrough and spurred very serious discussions about reproductive cloning of humans. Most people and certainly most nations and governing bodies agree that reproductive cloning of humans is not a good idea and should be banned. There are still many limitations in the technology that would make the likelihood of a normal human birth quite low. It is difficult to even successfully clone experimental animals, with close to 90 percent failure rate, and those that are cloned suffer from cancers, early deaths, and multiple abnormalities. Thus at this stage in human history, it appears that we should not be advocating for the production of populations of cloned humans. The process used for animals is called "somatic cell nuclear transfer" and offers tremendous possibilities. In this scenario, an entire somatic (body) cell including its nucleus and DNA, like from the skin or kidney, is placed into an enucleated egg cell. Stimulated with electric current or with the drug ionomycin, the new combination begins to divide, producing cloned cells under the direction of the genes from the transplanted cell. But scientists are very cautious about what to call this process, as reflected in the following excerpt from an article by Bert Vogelstein in the 2011 issue of *Science*:

> The goal of creating a nearly identical genetic copy of a human being is consistent with the term *human reproductive cloning*, but the goal of creating stem cells for *regenerative medicine* is not consistent with the term *therapeutic cloning*. The objective of the latter is not to create a copy of the potential tissue recipient, but rather to make tissue that is genetically compatible with that of the recipient. Although it may have been conceived as a simple term to help lay people distinguish two different applications of somatic cell nuclear transfer, "therapeutic cloning" is conceptually inaccurate and misleading, and should be abandoned.

In addition to the studies concerned with cloning are thousands of experiments with human and animal stem cells. These cells, by definition, have the potential to differentiate into any other cell in the body, and there are techniques through which scientists can recover adult stem cells from the bone marrow of animals and people. There are widely accepted methods that allow the collection of stem cells from animal and human embryos, and all of these exciting and rapidly developing technologies are focused on treating and curing diseases afflicting humans and animals and on understanding the fundamental mechanisms of biology, the study of life. And in 2012, the Nobel committee awarded the Nobel Prize in physiology or medicine to two of the pioneers of stem cell biology and cloning research. British scientist John Gurdon at the University of Cambridge and Japanese scientist Shinya Yamanaka at Kyoto University have provided phenomenal insights into the field of regenerative medicine that could allow the rebuilding of the human body with tissues generated from its own stem cells, a form of "regenerative medicine" discussed earlier.

Those supporting further pioneering in genetics point out that knowledge of heredity dates back to prehistoric times and that directing or controlling the traits of offspring has been done for centuries in the breeding of plants and animals. They point to the growing number of practical applications of genetic research and the tremendous promise of using genetics to cure human diseases. Rewriting the DNA misspelling that makes valine instead of glutamic acid in hemoglobin would cure sickle-cell anemia. Sim-ilarly, other potentially curable gene-based diseases or abnormalities include cancer, dwarfism, heart anomalies, Tay-Sachs disease, cystic fibrosis, cere-bral palsy, cleft palate, and mental retardation.

It is more than theoretically possible to rewrite the genetic message in the sperm or egg, thus raising the hope of preventing hemophilia and other diseases in the newborn of afflicted parents. In January 1996, a health baby boy was born to a Chinese couple in New York. The birth was remarkable because when the baby was a fetus he'd been diagnosed with alpha-thalassemia, a fatal genetic defect that occurs predominantly in Thai and Chinese people, in which the fetus is not able to make hemoglobin that carries oxygen in the red blood cells. This defect usually leads to a miscarriage within four months of conception, but the fetus and baby survived because he underwent transplants of cells in the bone marrow *while in the womb.*

By applying our minds and the methods of science, should we take advantage of the opportunities to enhance our ability to survive, to make life longer, richer, and healthier? Expanded applications of genetic engineering and experimental embryology, cloning, and transplants promise to be the new frontier of biological and medical developments in the twenty-first cen-tury. Further controversy is sure to ensue.

The Uniqueness of the Code Has Practical Applications

What Did They Have on O.J.?

In addition to the goals of the Human Genome Project, DNA analysis and alteration has myriad other direct positive impacts on our lives. One example was mentioned in Chapter 16—namely, the research being performed by drug companies to determine the particular DNA of the new strains of tuberculosis bacteria. In addition, because the genetic code that governs all life forms is identical in all species and because we now can read that code to stimulate the manufacture of proteins, an entire industry has emerged in the last twenty years—the biotechnology industry. The three-letter instructional sequences that make up a gene from one organism, such as a plant, can be placed in another organism, say a bacterium, and actually be read and followed by the latter. As a result, genetically altered bacteria have been developed and are now used by biotechnology companies to make human growth hormones and insulin, as described earlier. Understanding the central dogma (DNA makes RNA makes protein) has made it possible to help the body make critical protein molecules where it otherwise would lack that ability. As probably the best example of this concept, Amgen Inc. (named from its original inculcation as the American Genetics Co.) was founded in 1980 as a pioneer in the biotechnology revolution and still markets the two largest-selling products in the industry, Epogen and Neupogen. Epogen, synthesized from a human gene by bacteria, is a genetically engineered version of a protein called erythropoietin (EPO), which stimulates the production of red blood cells in bone marrow. The drug is used primarily by dialysis patients suffering from chronic anemia as a result of their failure to naturally produce sufficient amounts of EPO in their diseased kidneys. More than 175,000 patients in the United States alone use Epogen each year, eliminating their anemia and reducing the need for blood transfusions. The huge pharmaceutical company Johnson & Johnson also manufactures Epogen.

Neupogen, another of Amgen's major drugs, stimulates neutrophils, one of the body's infection-fighting white blood cells. It does so because, like Epogen, it's a genetically engineered protein identical to the naturally made proteins that stimulate neutrophils. Neupogen is used by cancer patients whose natural neutrophils were destroyed by chemotherapy. The drug not only wards off infections but allows patients to tolerate higher doses of chemotherapy over longer periods of time, which expands treatment options and enhances their chances of surviving cancer. Amgen is the world's largest independent biotechnology firm and has at least ten additional drugs in use or in clinical trials around the world. Like Amgen, other companies continue to pin their existence and future on the long-term viability of molecular biology–based drugs. Multiple companies like Johnson & Johnson, GlaxoSmithKline, AstraZeneca, and Pfizer are among those in the $65 billion per year pharmaceutical industry that is developing hundreds of new drugs annually. The Federal Drug Agency approves fifteen to twenty-five of these for use each year. Some will work and some won't; and a few will be miraculous lifesavers like insulin and penicillin. Other companies are using the ultimate molecular approach to assure the supply of corn and cotton crops so important throughout much of the world. *Bacillus thuringiensis* (*Bt*) is a bacterium that occurs naturally in the gut of some caterpillars. These bacteria produce crystalline proteins at certain times in their life cycle called endotoxins that were found to be poisonous to many insects, including flies and beetles that feed on corn plants. The gene that codes for the endotoxin was identified and sequenced, and in 1996, the first corn, potato, and cotton crops expressing the gene were grown and successfully protected from the insect pests by the bacterial endotoxin, newly expressed in the plants. By 2006, hundreds of thousands of acres of *Bt* crops had been planted worldwide, and the results appear to quite positive, though all opinions are not yet in. But the concept of inserting a known gene sequence with an established function into the genome of an animal or plant for the betterment of all humankind is an exciting direction to be heading.

Applying current knowledge of the genetic code, what did the Los Angeles police have on O. J. Simpson in his 1995 trial, and why has DNA become a more precise method than fingerprints to determine a person's identity? We've seen that the G-C, A-T codons are present in every living creature but are in different sequences for each living organism. Though there are only these four letters in the DNA alphabet, the long series of sentences, paragraphs, and volumes making up the genetic story within each of us is unique. With our newfound capability to read this story, it has become a powerful piece of evidence in identifying criminals who leave behind DNA in the form of blood, semen, hair, or skin. Thus through its expert witnesses the prosecution in the O. J. Simpson case introduced into evidence the unique sequences of Mr. Simpson's DNA from the blood found at the murder scene, in his infamous Ford Bronco, and at other locations.

Beginning in the mid-1980s, various tests were developed by molecular biologists for use in identifying individuals, not only in criminal cases but also in missing person cases and paternity suits. Unlike traditional hair testing, which determines only whether hair found at a crime scene has color and size characteristics consistent with the suspect's, or traditional blood and semen tests, which can pinpoint a suspect with 90 to 95 percent certainty, DNA typing can identify a suspect or other individual with virtual certainty. The statistics for the reliability of DNA typing are generally described in a range from 1 in 5 billion to 1 in 30 billion. This leaves the defendant with very narrow options to dispute this evidence because he or she can raise doubt only about the handling of the samples and whether the test was properly conducted. O. J. Simpson's attorneys were obviously successful in their extensive efforts to raise such doubts. Since the mid-1980s, DNA typing has gained general acceptance in the scientific and legal communities and has been used in thousands of murder, rape, burglary, and paternity cases.

The discovery of DNA has also led to great strides in the advancement of dentistry. Periodontal (gum) disease is one of the leading causes of tooth loss. Due to the advent of DNA probes, specific bacteria can be identified and eliminated with antibacterial therapy. The key to this process is to purify specific target DNA from several suspected bacterial pathogens. The DNA strand is then fragmented into small pieces and tagged with a radioactive label. The dentist then takes a sample of plaque from the patient's mouth and sends it to the laboratory for analysis. The technicians extract the bacterial DNA from the plaque sample and bind it to a filter. Similar DNA from the same bacteria will bind together when allowed to mix. Samples of DNA from the plaque are mixed with each of the radioactive bacteria DNA probes. If the suspected species of bacteria is present in the patient's sample, probe DNA will bind to it and can be viewed by exposing the sample to photographic film. The radioactive DNA will make a mark on the paper. The more bacterial DNA that is present, the darker the mark will be. The most prevalent bacteria in the plaque sample can be identified so that the proper antibiotics can be used to eliminate it. Thus in a process similar to identifying criminals' DNA, dentists are able to eliminate the "guilty" bacteria strains.

The discovery of the structure of DNA in 1953 was the closing of the circle that began with Darwin's subtle and powerful understatement in *On the Origin of Species* ninety-four years earlier: "Much light will be thrown on the origin of man and his history." Tying the origin of living things and evolution to the discovery of DNA's structure, Francis Crick stated:

> Every organism, every cell, and all the larger biochemical molecules are the end result of a long intricate process, often stretching back several billion years. . . . What is found in biology is *mechanisms*, mechanisms

built with chemical components and that are often modified by other, later mechanisms added to the earlier ones. . . . Nature could only build on what was already there.

Crick was astonished that in our modern culture so few people really understand natural selection. "They don't understand it," he surmised, "because the process is very slow, thus, we rarely have any direct experience of it operating," and there appears to be a contrast and contradiction "between the highly organized and intricate results of the process—all the living organisms we see around us—and the randomness at the heart of it." Said Crick, "some people also dislike the idea that natural selection has no foresight. . . . It is the 'environment' that provides the direction."

With the publication of *On the Origin of Species* in 1859, Charles Darwin took the first step toward fully understanding the basic internal mechanism that explains the origin of life and the ability of organisms to reproduce. A few years later, Gregor Mendel attributed heredity to invisible and unknown "heredity factors" in living things. Less than a century later, Crick and Watson answered the question, What is a gene?, and thereby became forever entwined with the legacies of Darwin and Mendel. Today, the Human Genome Project has unlocked the incredible story contained in our genes as our complex genetic ancestry unfolds.

EPILOGUE

In Focus

During the 1,000 long years of the Middle Ages, Western civilization was deeply entrenched in a state of mind that was hostile to rational thought. It would not have emerged from that mentality had it not been for a handful of European scholars who cleared a path that eventually led to the Renaissance. The remarkable progression of science began with the Copernican Revolution and marched through the nineteenth century, culminating with a great crescendo in the twentieth century, so far the most incredible century in the history of science.

In the brief span of the last 100 years our understanding of life and the universe has become rich and full. Like a photograph developing in the dark room, the image has become clear and focused. Our thoughts about the natural world have crystallized. This knowledge is embodied in the institution of science and has swept into every aspect of our daily lives. It is the foundation for the astounding technological achievements of this century.

In this epilogue we will reflect on the most incredible century, the complex institution that we call science, and the astonishing synthesis of formerly diverse disciplines that now characterizes modern science. We will touch on achievements in science as well failures, and on scientists' limited ability to direct or control how science and technology are applied in society. We will also briefly examine the importance of freedom of expression and the value of rational thought, without which science would not exist.

THE SYNTHESIS

There is one gift above all others that makes man unique among the animals . . . his immense pleasure in exercising and pushing forward his own skill. . . . Discovery is a double relation of analysis and synthesis together. As analysis, it probes for what is there. . . . As a synthesis, it puts the parts together in a form by which the creative mind transcends the bare skeleton that nature provides.
—Jacob Bronowski, *The Ascent of Man* (1973)

Seemingly unrelated events, diverse scientific discoveries, industrial trends, and religious outlook can all, in historical perspective, be observed to evolve . . . before they crystallize into a new pattern. . . . It is like looking into a chemical retort which is about to produce some rare and many-sided crystal. One moment everything is in solution . . . and yet in the next instant a shape has appeared out of nowhere.
—Loren Eiseley, *Darwin's Century* (1961)

We entered the twentieth century riding horses. We have left it riding space-ships. We entered the century dying of typhoid and smallpox, and now in the twenty-first century we've conquered those diseases and numerous others. At the turn of the nineteenth century, organ transplants were unthinkable, while in this century many survive because another person's heart or other donated vital organ sustains them. In 1900 the average human life span was forty-seven years. Today it is seventy-five. We entered the twentieth century communicating over short distances with the newly invented wireless radio. We now send voices and color pictures back and forth across billions of miles of space.

We entered the twentieth century without any idea of the dimensions of the universe, believing that nothing lay beyond the Milky Way Galaxy and that the stars are motionless. We left the century with the knowledge that we ride a tiny sphere near a star on the far reaches of a rotating galaxy among billions of other stars and galaxies, all still reeling from a violent explosion that took place 14 billion years ago.

We entered the twentieth century satisfied with Newton's physics and now know that Einstein's principles are needed to fully explain the universe. As we entered the twentieth century, we had just begun to explore Leucippus's atom. As the twenty-first century develops, we've broken the atom into particles and proved the existence of the Higgs boson that imparts mass to these particles. It is with astonishment that we realize that one nuclear weapon can release more destructive power than all the energy of all the weapons in all the battles in the previous history of humankind.

In 1900, we suspected Darwin was right. As we moved well into the next millennium, we removed all doubt as we have unearthed Lucy, "the Handy Man," and others of our ancestors. We have come to understand the intricate workings of the cell as we further manipulate the DNA molecule and genetic codes and finally clone cells, tissues, and animals, all to our ultimate benefit.

Reason Becomes the Soul of Science

Myths in the Hands of Demagogues

In 1560, Giambattista della Porta, an Italian physicist, established the world's first organization dedicated specifically to the exchange of ideas among scientists. His institution, Academia Secretorum Naturae (Academy of the Mysteries of Nature), was controversial and short lived but signaled the dawn of the Age of Discovery in the Renaissance. In 1620, Francis Bacon (1561–1626) wrote *Novum Organum*, which was a modern version of Aristotle's *Organum*. Bacon, an English philosopher, member of Parliament, and chancellor of England under King James I, was the first to formalize detailed rules for ensuring truth in the pursuit of science. He is credited with establishing what is now universally accepted as the scientific method. Galileo, a contemporary of Bacon, further developed and refined it.

During the Renaissance we began to recognize and accept what the scientists were finding. Aristotle's and the church's view of changelessness was abandoned—the planets and stars are not motionless. Astronomers like Tycho Brahe and Galileo made their observations and developed hypotheses. With the advent of the scientific method, new observations developed into hypotheses that survived scrutiny and became theories. Over time, theories that survived efforts to disprove them became laws and principles, for they must be repeatable and have the ability to survive rigorous and inevitable challenge.

In the seventeenth and eighteenth centuries, the governments of Europe began to provide extensive financial support for scientific research. After Galileo's trial, it became common in Italy, England, France, and Germany to advocate science, experimentation, and theories despite the opposition of religious leaders in those countries. Thus science had not only developed institutions and a method by which to work but had finally gained a safe haven—a "zone of freedom" in which to grow and prosper.

In the twenty-first century, we've now immersed ourselves in what Newton called the vast sea of knowledge. We've felt and tasted the water. We've washed it over ourselves, and what we've learned would astound him and the other great and imaginative minds of generations past. Science and its tradition of reason, applied properly by governments in a moral world society, must remain one of humankind's foremost goals. Again, as Jacob Bronowski said, "There is one gift above all others that makes man unique among the animals . . . his immense pleasure in exercising and pushing forward his own skill." Discoveries in science have opened our minds and justified our faith in our unique abilities.

Understanding the universe and ourselves must continue to be the goal of science. In order to achieve that goal, institutions must exist that best facilitate a free and prosperous society. We must always be cognizant of how fleeting that freedom can be. Human advancement in all respects is bound up tightly with freedom. Intellectual growth is stifled when freedom of action or expression bring threat of sanction. When change in long-held beliefs is suggested, the response of many people is to restrict the free exchange of ideas that support such change. Thus lack of freedom and fear of change are closely related. Insistence on the status quo is often based on myth, not a conscious suppression of truth, yet this can be more of a threat to freedom than suppression fueled by outright lies. As John F. Kennedy said at the Yale commencement exercises in 1962, "The great enemy of truth is very often not the lie—deliberate, contrived and dishonest—but the myth—persistent, persuasive and realistic." Myths continue to proliferate today and in many respects dominate thought throughout the world.

Freedom does not depend on the law but requires the public to understand and appreciate rational thought. Reason is the soul of science. Once it is suppressed or abandoned, totalitarianism or anarchy quickly fill the void and impact the entire community, not only science. For example, if a society abandons rational thought and allows or mandates that creationism be taught in public schools, where would it stop? People for the American Way, a nonpartisan constitutional liberties organization based in Washington, DC, issues an annual survey of censorship and related challenges to public education. Its recent reports state that there is "a renewed fight to add Creationism to science classes" and that would-be censors of teaching evolution "are increasingly using the state legislatures in an attempt to impose their ideological . . . agendas on the nation's classrooms." In some polls taken across populations in the United States, more than 80 percent of respondents thought that evolution should be taught in public schools while almost half of these same folks believe there is a place for teaching creationism, but that it should be classified as a religious belief, not a science. Numerous legislative initiatives that failed would have made teaching evolution a basis for dismissal. Support for including creationism in public school curricula is increasing as is the likelihood of this type of bill becoming law in the near future. A society that maintains such irrational laws could readily condone and support other destructive laws and social policies, such as those regarding race, color, and religion. Intellectual, artistic, and scientific progress would cease, as is the case in some totalitarian societies that exist today.

After escaping from the oppressive Soviet Union, nuclear physicist and human rights activist Andrei Sakharov (1921–1989) wrote that "Intellectual freedom is essential to human society. . . . Freedom of thought is the only guarantee against an infection of people by mass myths, which, in the hands of treacherous hypocrites and demagogues, can be transformed into bloody dictatorships." Mentalities and views of the world can't be changed by decree. They can be formed or changed only by open discourse in a society that is not afraid of discarding old ways of thinking and is interested in expanding its horizons. Periods of severe antirationalism have punctuated human societies through all recorded millennia. This mentality ebbs and flows throughout history and is remarkably alive and intense today, making it as imperative now as at any time in history to cherish and support freedom of expression and our reliance on reason and rational thought.

The Progress and Application of Science Is Controlled by Our Political and Social Institutions

Glowing Blue Powder and 14 Billion People

In Chapter 19, we examined controversy surrounding the human genome and genetic engineering. Debate also continues on the use of nuclear power and the atomic bomb, and the application of other discoveries and technology. Science and scientists are often blamed for negative impacts of such discoveries. Yet the decisions concerning the use of such knowledge are made by society as a whole primarily through governmental institutions. This includes the debate over allocation of resources. In many nations, those who want products of technology for combat and destruction compete with those who seek funds for healthcare and other social needs. Some of those countries have the best jet fighters, but many of their people have no toilets, no bath, no furniture, no medicines, and no education. Despite our great advances in science, much of humanity lives in ignorance and poverty. But as Einstein wrote, "Rational thinking does not suffice to solve the problems of our social life."

More than 400 children die every hour from diseases such as cholera, typhoid fever, dysentery, and hepatitis. Every year, millions of people, mostly children, die of starvation. Where are the blessings of technical civilization and the progress of science where these children lived?

It is not only in war that scientific discoveries and technological advances have played a role in death and destruction. On December 3, 1984, a release of the toxic gas methyl-isocyanide from a plant in Bhopal, India, owned by

Union Carbide Corporation eventually caused the death of over 3,000 people. On April 26, 1986, a reactor at the Chernobyl Nuclear Power Station in Soviet Ukraine spewed a radioactive cloud across a 10,000-square-mile area of Europe, exposing 5 million people, and releasing 200 times as much radiation as the atomic bombs dropped on Hiroshima and Nagasaki in 1945. Thirty-two people died immediately and hundreds of thousands more came down with radiation-related illnesses, with some estimates stating that as many as 150,000 eventually died from such diseases. In the shadow of the concrete and steel sarcophagus that now entombs reactor no. 4, the once-thriving Soviet "model city" of 40,000 people—Pripyat—is empty. About 270,000 people still live in dangerously contaminated areas.

In 1987, at an abandoned radiology clinic in Goiânia, Brazil, a group of curious youngsters broke open a 300-pound lead capsule containing cesium-137, a radioactive substance used in cancer treatment. The glowing blue powder inside the capsule eventually contaminated 249 people, resulting in several deaths, amputations, and related illnesses.

In the largest nuclear disaster since Chernobyl, the Fukushima I Nuclear Power Plant, located in Okuma, Fukushima Prefecture, Japan, experienced equipment failure, nuclear meltdown, and release of radioactive materials. This was the result of an earthquake and consequent tidal wave on March 11, 2011. This situation was rated a 7, the highest level of disaster given by the International Nuclear Event Scale. In July 2012, a report analyzing the disaster and its likely long-term impact was submitted to the Japanese government, but only time will reveal the fate of the thousands of individuals exposed to varying levels of radioactive materials.

The ultimate and most troubling paradox imaginable concerns the great progress of medical science. While it has saved so many lives, it has indirectly resulted in a population increase that now threatens the world. Look at these startling facts:

- In 1850, the world population was 1.2 billion.

- In 2010, the world population was more than 6.8 billion.

- One-half of all people who have ever been born are alive today.

- In the next *fifteen years*, the world population will increase by more than the total number of people (1.2 billion) who inhabited the planet in 1850.

As mentioned earlier, the average human life span has increased from forty-seven years in 1900 to seventy-five years today. *Thus in the equivalent of two lifetimes, the world population has increased more than fivefold!* By reducing additional major risk factors, such as high blood pressure, obesity, and smoking, the average life expectancy could increase to ninety-eight or ninety-nine years. As the elderly population increases, more people are likely to suffer from such long-term degenerative diseases as blindness, arthritis, Alzheimer's disease, sensory impairment, osteoporosis, and other disorders, placing greater strain on every nation's economy. The world population is increasing by 3 people every second. Each year, the world population increases by 90 to 100 million people—about the combined population of the Philippines and South Korea—with the largest increases in the poorest countries, which are the least equipped to meet the needs of the new arrivals and invest in their future. Experts warn that unless the trend is reversed soon, the world's population will reach 14 billion people in a hundred more years.

Rapid population growth damages economic gains many countries have achieved, holding back or crippling altogether the prospects of balanced development. It's not the absolute size that a country achieves, it is the rate of change that prevents them from maintaining agricultural productivity, creating new jobs, delivering healthcare, and developing the other advances needed. With the exploding world population, there is greater potential for violence and social upheaval. Thus birth control and family planning are critical elements in achieving world peace. As a result of complicated social and political factors, half the world's women have no access to family planning services. Yet, women are the audience that must be reached and educated if developing nations are to successfully stem population growth. H. G. Wells once said that civilization is engaged in a race between education and catastrophe. Catastrophe in the form of runaway population growth is winning.

However, the scientists' role in the world's social ills is only indirect. The solution is not in science. Our failings are not in science. They are found in the social and cultural structures and institutions we've created because politics, government, and our own morality determine how science and technology will be *applied*. Einstein noted that "science can only ascertain what *is*, not what should be. . . . Outside of its domain, value judgments of all kinds remain necessary." Robert Oppenheimer, reflecting on his own role in the decision to drop the atom bomb, pointed out that as director of the Manhattan Project, "I had no policy-making authority." Similarly, the scientists throughout the Human Genome Project are working on scientific advances, not social policy.

The object of science and technology is not to conquer nature or to conquer people. For Kepler and Newton, it was to understand how the universe works. For Watson and Crick it was to understand how life works. Marie Curie called science the most sublime manifestation of the human spirit and a perfect tool to aid humanity. As an institution, science seeks primarily to comprehend the physical world in a logical and uniform system of thought.

Diverse Fields of Science Merged in the Twentieth Century

Fertility of Hybrid Species

Loren Eiseley was referring to the history of evolutionary theory in the passage quoted at the beginning of this epilogue. But his statement about diverse scientific discoveries crystallizing into a new pattern is equally applicable to all of science in the twentieth century. In the last century, all the branches and fields of science coalesced into one overlapping, interwoven, intertwined, and interdependent complexity of understanding that is now hurling us into the future. The borderlines that used to clearly separate astronomy, physics, and biology have become fertile fields of common interest and pursuit, together with other fields and subdisciplines. As we saw in Part Seven on DNA, twentieth-century physicists turned to biology, and biologists and geneticists found their answers in chemistry. In the twentieth century, astronomers became dependent on physics and chemistry, while geology revealed key aspects of the origin and evolution of life. "One moment everything was in solution . . . ," wrote Eiseley, "in the next instant a shape has appeared out of nowhere."

Today, scientists are primarily experimenters, not theorists—they are molecular geneticists, biochemists, biophysicists, cosmologists, and other *combinations* of expertise not fathomed in earlier centuries. In the 1500s, Copernicus made his twenty-seven observations and then applied logic to arrive at heliocentricity. Similarly, Newton, Darwin, and Einstein epitomize the lone scientist who amassed the evidence and mulled and formulated it until the theories were shaped. Four of the five great discoveries made in the twentieth century—structure of the atom, the nature of the Big Bang, cell structure, and DNA—were made possible because "one-person science" had ended and "group science" is upon us. In the twentieth century, science became symbiotic and overlapping and scientists no longer desired to maintain their isolation within a defined field. They recognized the need for cross-fertilization—the need to follow developments in the other fields on which they depended and to share information.

We saw the first signs of the connection between astronomy and physics in the early 1600s when Kepler discovered his Laws of Planetary Motion. The final synthesis of those fields occurred in the twentieth century, as the Big Bang theory took shape. In the first decades of the past century, the fields of biophysics, biochemistry, and molecular genetics developed. Beginning in 1936, George Gamow developed the first theory linking small-scale nuclear processes to the Big Bang. Biologists, chemists, and geneticists have pooled their knowledge in myriad ways that will remain linked forever. Carbon, nitrogen, oxygen, and phosphorus are the principal building blocks of life. Protein, which is essential to all organic life, depends on the capture of nitrogen from the atmosphere. Nitrogen is also an essential component of DNA. Further examples touch many of the other elements—sodium and potassium are critical elements of the nervous system and brain, and calcium is the most essential component in maintenance of the form of organisms in bones and shells.

In his famous book *A Brief History of Time*, physicist Stephen Hawking tied atomic physics to biology, explaining that if the electric charge of the electron had been only slightly different, either the nuclear fusion process that transforms hydrogen into helium in stars would not have been possible, or many stars wouldn't have exploded. In either case, without sunlight or the heavier chemical elements made in stars and flung into space when stars explode, life as we know it on Earth could not have developed.

In reminiscing on the history of classical genetics, Francis Crick observed that "the important thing was to combine it with biochemistry." In his book *What Mad Pursuit*, in which he discusses the process of scientific discovery, Crick describes geneticists' and protein chemists' belated recognition of their need to become more familiar with each other's field to properly conduct experiments designed to solve the coding problem of the DNA molecule. As late as 1954 (after the discovery of the double helix), experts in genetics remained extremely hesitant to spend time learning about protein chemistry, and vice versa. As discussed in Chapter 19, the chemist Vernon Ingram then identified a specific alteration in the code for one amino acid in the protein hemoglobin as being responsible for sickle-cell anemia. Crick credited Ingram's research with "a drastic change of attitude" of the geneticists and protein chemists, and noted the lesson to be learned about the synthesis of diverse scientific disciplines. "In nature," stated Crick, "hybrid species are usually sterile, but in science the reverse is often true. Hybrid subjects are often astonishingly fertile. . . . If a scientific discipline remains too pure it usually wilts."

Twentieth-Century Science Led to the Science of Incremental Advances

Anonymity and the Nobel

The refinement and greater understanding of scientific principles has led to much greater specialization among scientists. Each new issue, each scientific question posed, each theory, now depends on the cross-fertilization of ideas and research among diverse fields of science by scores, hundreds, or even thousands of scientists driving toward a collective creation, with each of those people concentrating on his or her narrow specialty. Thus, contrary to the seven great and broad sweeping theories that have revolutionized science, as presented in this book, the latter years of twentieth-century science through today have seen the science of *incremental advances*—that is, narrow and complicated discoveries derived from experiments and studies in research laboratories all over the world, all contributing to an ultimate understanding of some limited phase of the physical universe or living organisms. As Louis Pasteur once said, "Science advances through tentative answers to a series of more and more subtle questions which reach deeper and deeper into the essence of natural phenomena."

The emphasis in science today is on research and experiments that refine discrete aspects of larger, already proven discoveries. Also, it is probable there are few (if any) scientific discoveries remaining that would be comparable in significance to the seven reviewed in this book. In *The End of Science,* John Horgan predicts that while "this is the age in which we are discovering the fundamental laws of nature . . . that day will never come again." Twenty-first-century science has become the science of the infinitesimal, rather than the science of the infinite. Even the science of all matter in the universe—the Big Bang—has become the science of subatomic particles, atom smashers and particle accelerators, as scientists search for the Grand Unified Theory that seeks to tie quantum theory to the cosmos. This is the theory for which Albert Einstein searched in vain for forty years, one that would combine chemistry, electromagnetism, mechanics, astrophysics, and the rest of physics into one set of equations that explains *everything.*

By the last half of the twentieth century, science had become that of the Hubble Space Telescope, the electron microscope, and molecular biology. It is the science of privately funded laboratories, and thousands of university and government laboratories engaged in fundamental research, not the science of a naturalist collecting specimens of finches in the Galapagos Islands. As examples of such facilities, Brookhaven National Laboratory in Upton, Long Island, has a 5,000-acre campus where over 3,000 people conduct research in the most esoteric fields of particle physics, centered around the Alternating Gradient Synchrotron, a particle accelerator that was used in three of the lab's seven Nobel prizes in physics and chemistry. The Research Triangle Park (RTP) that sits among Raleigh, Durham, and Chapel Hill, North Carolina, is the home of the National Institute of Environmental Health Sciences (part of the U.S. Department of Health and Human Services) and employs about 300 people on its campus that spreads over the wooded hills. Among the research carried out there are efforts to determine how chemicals in our environment cause cancer and miscarriages and how expression of certain genes leads to a variety of cancers. RTP is surrounded by North Carolina State University in Raleigh, the University of North Carolina in Chapel Hill, and Duke University in Durham. These three outstanding universities are world leaders in engineering, toxicology, and molecular medicine, and similar concentrations of outstanding science can be found from coast to coast across the United States. James Watson worked at the Cold Spring Harbor Laboratories in New York and was its director for thirty-five years. Fundamental processes governing gene expression, genomics, and bioinformatics are being explored at this beautiful and highly productive facility. Eight Nobel laureates have worked at the laboratory, which currently employs around 400 scientists. Similar levels of scientific achievement are found worldwide at such well-known institutions as the Pasteur Institute in Paris, the Max Planck Institute with numerous research centers in Germany, Oxford and Cambridge Universities in the United Kingdom, and well-developed universities and research entities in Mexico, Argentina, Brazil, and Chile.

Twenty-first-century science interweaves individual contributions on a worldwide scale into the science of the multitudes, rather than the science of the one. Breakthroughs and discoveries are made by groups, teams, laboratories, and universities, rather than individuals. Even the naming of new elements added to the periodic table, which used to be the privilege of the discoverer, is being usurped by a committee of the International Union for Pure and Applied Chemistry (see Chapter 4).

Despite the transition to group science, individual achievements are recognized and rewarded in many ways, particularly through the Nobel Prize, the crowning achievement in fundamental science. The Nobel Prize, first awarded in 1901, is paid from a fund established in 1896 by the will of the Swedish chemist and engineer Alfred Bernhard Nobel (1833–1896), who invented dynamite. Nobel's will specified that the awards should be made annually "to those who during the preceding year, shall have conferred the greatest benefit on mankind" in three fields of science: physics, chemistry, and physiology or medicine. Nobel Prizes are also awarded annually in literature, peace, and economics.

FIGURE E-1

The Six Levels of Scientific Achievement

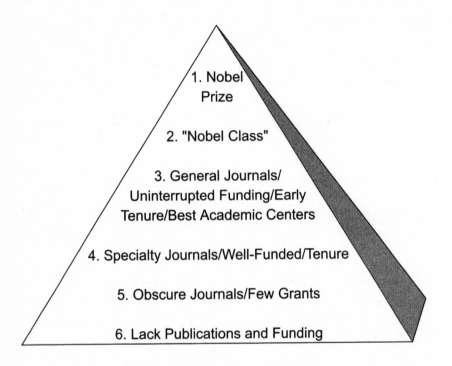

The pyramid in Figure E-1 depicts the authors' view of the multiple levels of scientific achievement within the scientific community. At the pyramid's apex are the winners of the Nobel Prize, people like Curie, Einstein, Crick, Watson, and Rodbell. In the second tier are the "Nobel Class"—those whose work is just as imaginative, groundbreaking, and important to humanity as the Nobel winners, but who weren't the first to make a particular scientific breakthrough. Level three is occupied by those who publish solely in the best general science journals, such as *Science, Nature,* and *Cell.* They enjoy uninterrupted funding from federal government agencies like the National Institutes of Health and the National Science Foundation, and from private foundations like the Howard Hughes Medical Institute. Level four scientists are well known in their specialties, have made significant advances, are tenured at prominent universities, are regularly funded in support of their work, and publish routinely in the best journals of their fields of study and occasionally in the most prestigious general journals. The fifth level publishes, but usually in obscure and sparsely read journals. They haven't made an impact in their fields and have difficulty attracting grant money to support their pursuits. Finally, the individuals on the sixth and bottom rung of scientific pursuit rarely produce a noteworthy paper or receive a grant. They begin as postdoctorates or junior faculty, where essentially all scientists begin, but remain mired on the bottom while others ascend to the higher levels of scientific achievement at noteworthy institutions. The people in the pyramid number in the tens of thousands around the world, with fewer individuals in each ascending level, and are the people making the incremental advances in science.

The existence of group science, while still recognizing the value of individual achievement, has resulted in a paradox. Despite the difficulty of their achievements, even the greatest twentieth- and twenty-first-century scientists—the ones on the top levels of the pyramid—are anonymous and largely unknown to the general public. As we saw in the preceding chapters, Marie Curie, Ernest Rutherford, Albert Einstein, Niels Bohr, Walther Flemming, Edwin Hubble, Thomas Morgan, Francis Crick, James Watson, Arno Penzias, and Robert W. Wilson were all giants in their fields. Indeed, Watson, Penzias, and Wilson are still active in their fields. But in the average household, the people on the long list of Nobel Prize recipients since 1901 are hardly known, except for Einstein and perhaps Curie.

The last of the seven great scientific discoveries was made in 1953 by Crick and Watson. *Are there no great scientists from the latter half of the twentieth century and today?* Yes, there are great scientists, but the opportunities for mass recognition have diminished because their work, measured in incremental advances, is so complex. Though Nobel Prizes are given to individuals each year, the incremental advance for which each of those people receives the prize will continue to be consumed by the intricate labyrinth of the science of the last 115 years. Finally, in 2012, Bruce Alberts, editor in chief of *Science* magazine, asks in the September issue if we have seen "The End of Small Science." He is quick to answer no but makes the important point that some facets of modern science will remain as "big science," such as ENCODE, a decade-long project involving more than 400 international scientists who have compiled the Encyclopedia of DNA Elements (ENCODE) that lists the known functional elements of the human genome. This information is central to further research on fundamentals of biology, health, and disease. The huge international arrays of telescopes discussed in Part Four should be included in big science. But the small-science era, the incremental advances as described earlier, is far from over.

The Seven Greatest Scientific Discoveries in History . . .

While we continue to argue over political structures and religious philosophies, and while we are not sure how best to manufacture and distribute the world's goods or allocate wealth among people, we now know a great deal about what *exists*. We find ourselves with more revealed than could have been conceived in our wildest dreams of just a few centuries past.

All the matter in the universe still reels from an explosion that occurred 13 to 15 billion years ago. During the first 10.5 billion years of the universe's existence, this planet coalesced out of dust and gases, then gave birth to the first cell. It took over 3 billion years for that cell to evolve into the first multicelled organisms. Plants evolved that swallow photons from the sun, mix them with water and carbon dioxide into a carbohydrate brew, while the animals inhale the plants' waste—oxygen—and return their own discarded molecules, the carbon dioxide so vital to the plants' survival. All the living organisms on this journey are the natural products of this rotating and changing sphere that revolves around an ordinary star 93 million miles away in an outer spiral arm of the Milky Way Galaxy, amid billions of other galaxies.

Another few hundred million years elapsed between the appearance of the first mammal and *Australopithecus*. During the next 3 million years, *Australopithecus* evolved by natural selection from a figure in the landscape into *Homo sapiens*, shaper of the landscape, a being intelligent enough to wonder at the universe and understand that we are 6 billion individuals made up of faint, elusive, and finely adjusted electrical charges, configured into 60 trillion cells and pulled down firmly to this Earth by the interplay of those charges.

. . . and the People Who Made Them

After standing on the shoulders of Kepler, Galileo, and others, Newton led humankind to the edge of the vast sea of knowledge. Ernest Rutherford, Marie Curie, and Niels Bohr discovered Leucippus's ultimate particle. Albert Einstein identified the enormous power hidden within that particle, and redefined time and space. Edwin Hubble peered through the telescope and saw the galaxies blowing apart. Charles Darwin looked at the vast diversity of plant and animal life and recognized a common ancestor. Through the microscope, Schleiden, Schwann, Flemming, and Weismann came to the realization that they were viewing the source of all life and growth. In generations of peas, Gregor Mendel saw undetectable heredity factors being passed on through some unknown process. East, Morgan, and others identified that process. Francis Crick and James Watson opened the secret text of life, the first words of which were written in the first cell, 4 billion years ago.

Knowledge is our destiny. *Homo sapiens* will continue to search for the answers to new questions. We will develop new concepts, new theories, and we will continue our quest to understand the natural world. We are different from the other animals. We must continue to discover, create, explore, and invent. We must search for the cure and the lifesaving solution, for we are the discoverers, creators, explorers, and inventors. We seek the unknown—the deep, the dark, the never-before seen—and we have within us the capacity for even greater wisdom.

We have come to the future. We have found our place by looking back and understanding history. We have already become twenty-first centurions. As Orville Wright stated, "We don't have to look too far to see the future. We can already see it will be magnificent."

We have now traveled the 14-billion-year journey.

Chronology of the Seven Greatest Scientific Discoveries in History

	BEFORE 1800		1800	
1. GRAVITY/ PHYSICS	Copernicus's *The Revolutions* (1543) Kepler's Laws of Planetary Motion (1609–1618)	Galileo's *Dialogue on the Two Chief World Systems* (1632) Newton's *Principia* (1687)		
2. ATOM		Boyle's Skeptical Chymist (1661)	Lavoisier identifies oxygen (1783) Dalton's atomic theory (1803)	Mendeleev's Periodic Table (1869)
3. RELATIVITY				Maxwell's four laws/electromagnetic field (1861)
4. BIG BANG		Herschel discovers Milky Way (1785)		Maxwell's four laws/electromagnetic field (1861)
5. EVOLUTION		Linnaeus' *Systema* (1735) Hutton founds geology (1785)	Lyell's Principles of Geology (1830)	Darwin's *Origin of Species* (1859) Darwin's *Descent of Man* (1871)
6. CELL/ GENETICS		Van Leeuwenhoek views blood cells (1673)	Brown observes cell nucleus (1831) Schleiden and Schwann develop cell theory (1838–39) Mendel's principles of heredity (1865)	Virchow's *Cellular Pathology* (1858)
7. DNA				

1900

2015

Röntgen discovers X-rays (1895)

Einstein's equation $E = mc^2$ (1907)

Chadwick discovers neutron (1932)

Meitner describes chain reaction (1939)

Thomson discovers Electrons (1897)

Rutherford describes atom (1912)

Cockcroft/Walton achieve fission (1932)

Fermi achieves large-scale chain reaction (1942)

Curie discovers Radioactivity (1898)

Bohr develops quantum theory (1913)

Einstein's letter to FDR (1939)

Atomic bombs on Hiroshima and Nagasaki (1945)

Higgs boson (2012)

Einstein's Special Theory of Relativity (1905)

Einstein's General Theory of Relativity (1915)

Hubble determines Andromeda is galaxy (1923)

Penzias and Wilson discover cosmic background radiation (1965)

Continuing expansion of the universe (2011)

Hubble discovers recession of galaxies (1927)

Johanson discovers "Lucy" (1974)

New human ancestors Ardipithecus ramidus (2009) and Australopithecus sediba (2011)

Discovery of hominid fossils (1920–present)

Discovery of seafloor spreading and plate tectonics (1963)

Flemming discovers chromosomes (1879)

Mendel's principles rediscovered (1900)

Discovery of organic molecules on Mars (1996 and 2003)

Weismann describes meiosis (1886)

Morgan discovers recombination and identifies genes (1915)

Fisher, Haldane, Wright develop population genetics (1920)

Discovery of exoplanets (2011)

Levene identifies nucleic acids (1909)

Wilkins, Franklin x-ray diffraction (1945–1953)

Sanger sequences RNA (1960)

Sequencing of the Human Genome (2001)

Avery proves DNA contains genes (1944)

Crick, Watson discover DNA structure (1953)

Neanderthal DNA found in the modern human genome (2009)

Chargaff determines A–T/C–G ratios (1950)

Human Genome Project begins (1989)

BIBLIOGRAPHY

PART ONE

Gravity and the Basic Laws of Physics

Andrade, E.N. da C. *Sir Isaac Newton* (New York: Doubleday, 1958).

Augros, Robert M., and George N. Stansciu. *The New Story of Science* (New York: Bantam Books, 1984).

Baumgardt, Carola. *Johannes Kepler, Life and Letters* (New York: Philosophical Library, 1951).

Bhattacharjee, Yudhijit. "A Distant Glimpse of Alien Life?" *Science* 333 (2011): 932.

Caspar, Max. *Kepler* (New York: Abelard-Schuman, Ltd., 1959).

Christianson, Gale E. *In the Presence of the Creator* (New York: The Free Press, 1984).

Clarke, Arthur C. *The Promise of Space* (New York: Berkley Books, 1985).

Cohen, I. Bernard. *Introduction to Newton's Principia* (Cambridge, MA: Harvard University Press, 1978).

Cooper, Lane. *Aristotle, Galileo, and the Leaning Tower of Pisa* (New York: Ithaca Press, 1935).

Drake, Stillman. *Discoveries and Opinions of Galileo* (New York: Doubleday Anchor Books, 1957).

Drake, Stillman. *Galileo: Pioneer Scientist* (Toronto: University of Toronto Press, 1990).

Drake, Stillman. *Galileo Studies* (Ann Arbor, MI: The University of Michigan Press, 1970).

Drake, Stillman. *Galileo at Work* (Chicago: The University of Chicago Press, 1978)

Ferris, Timothy, ed. *The World Treasury of Physics, Astronomy, and Mathematics* (Boston: Little, Brown and Company, 1991).

Heilbron, J. L. *Galileo* (New York: Oxford University Press, 2010).

"Infinitely Fast Light." *Science* 338 (2012): 727.

Koestler, Arthur. *The Watershed* (Lanham, MD: University Press of America, 1960).

Koyre, Alexandre. *Newtonian Studies* (Cambridge, MA: Harvard University Press, 1965).

Motz, Lloyd, and Jefferson Hane Weaver. *The Concept of Science* (New York: Plenum Press, 1988).

Sagan, Carl. *Cosmos* (New York: Random House, 1980).

Sagan, Carl. *Murmurs of Earth* (New York: Ballantine Books, 1978).

Service, Robert F., and Adrian Cho. "Strange New Tricks with Light." *Science* 330 (2010): 1622.

Thoren, Victor E. *The Lord of Uraniborg* (New York: Cambridge University Press, 1990).

Westfall, Richard S. *The Life of Isaac Newton* (New York: Cambridge University Press, 1993).

Ziegler, Philip. *The Black Death* (Wolfeboro Falls, NH: Alan Sutton Publishing Inc., 1991).

PART TWO

The Structure of the Atom

Asimov, Isaac, *Understanding Physics: The Electron, Proton, and Neutron* (New York: New American Library, 1966).

Atkins, P. W. *The Periodic Kingdom* (New York: Basic Books, 1995).

Brady, James E., and John R. Holum. *Fundamentals of Chemistry* (New York: John Wiley & Sons, 1984).

Cho, Adrian. "Higgs Boson Makes Its Debut After Decades-Long Search." *Science* 337 (2012): 141.

Crease, Robert P., and Charles C. Mann. *The Second Creation* (New York: Macmillan Publishing Company, 1986).

Curie, Eve. *Madame Curie* (New York: Doubleday & Co., 1937).

Davies, Paul. *God and the New Physics* (New York: Simon & Schuster, 1983).

Feinberg, J. G. *The Story of Atomic Theory and Atomic Energy* (New York: Dover Publications, 1960).

Feynman, Richard P. *"Surely You're Joking, Mr. Feynman!"* (New York: Bantam Books, 1986).

Gamow, George. *Atomic Energy in Cosmic and Human Life* (New York: Macmillan, 1947).

Holden, Alan. *The Nature of Atoms* (New York: Oxford University Press, 1971).

Ihde, Aaron J. *The Development of Modern Chemistry* (New York: Dover Publications, 1984).

Leal, J. P. "Chemical Elements: What's in a Name?" *Science* 334 (2011): 176.

Malone, Leo J. *Basic Concepts of Chemistry* (New York: John Wiley & Sons, 1981).

Morris, Richard. *The Edges of Science* (New York: Prentice Hall Press, 1990).

Nitske, W. Rob. *Wilhelm Conrad Röntgen* (Tucson: The University of Arizona Press, 1971).

Packard, Edward. *Imagining the Universe* (New York: Perigee Books, 1994).

Pflaum, Rosalynd. *Grand Obsession: Madame Curie and Her World* (New York: Doubleday, 1989).

PART THREE

The Principle of Relativity

"At Long Last, Gravity Probe B Satellite Proves Einstein Right." *Science* 332 (2011): 649.

Bondi, Hermann. *Relativity and Common Sense* (New York: Dover Publications, 1964).

Clark, Ronald W. *Einstein: The Life and Times* (London: Hodder and Stoughton, 1973).

Einstein, Albert. *Ideas and Opinions* (New York: Crown Publishers, 1954).

Einstein, Albert. *The World As I See It* (New York: Philosophical Library, 1956).

Einstein, Albert. and Leopold Infeld. *The Evolution of Physics* (New York: Simon and Schuster, 1938).

Frank, Philipp. *Einstein, His Life and Times* (New York: Alfred A. Knopf, 1947).

Gabor, Nathaniel M., Justin C. W. Song, Qiong Ma, et al. "Hot Carrier–Assisted Intrinsic Photoresponse in Graphene." *Science* 332 (2011): 649.

Hawking, Stephen. *Black Holes and Baby Universes* (New York: Bantam Books, 1993).

Hawking, Stephen W. *A Brief History of Time* (New York: Bantam Books, 1988).

Jeans, James. *Physics and Philosophy* (New York: Dover Publications, 1981).

Kern, Stephen. *The Culture of Time and Space 1880–1918* (Cambridge, MA: Harvard University Press, 1983).

Michelmore, Peter. *Einstein, Profile of the Man* (New York: Dodd, Mead & Co., 1962).

Pais, Abraham. *Subtle Is the Lord: The Science and the Life of Albert Einstein* (New York: Oxford University Press, 1982).

Russell, Bertrand. *The ABC of Relativity* (New York: New American Library, 1985).

Sartori, Leo. *Understanding Relativity* (Berkeley: University of California Press, 1996).

Sayen, Jamie. *Einstein in America* (New York: Crown Publishers, 1985).

Schilpp, Paul Arthur, ed. *Albert Einstein: Philospher-Scientist* (New York: The Library of Living Philosophers, 1949).

Swenson, Lloyd S., Jr. *Genesis of Relativity* (New York: Burt Franklin & Co., 1979).

PART FOUR

The Big Bang and the Formation of the Universe

Asimov, Isaac. *The Universe* (New York: Avon Books, 1966).

Barrow, John D. *The Origin of the Universe* (New York: Basic Books, 1994).

Cruz, M. "Inescapable Pull: Black Holes." *Science* 337 (2012): 535.

Davies, Paul. *The Last Three Minutes* (New York: Basic Books, 1994).

Fall, S. M., and D. Lynden-Bell, eds. *The Structure and Evolution of Normal Galaxies* (Cambridge, England: Cambridge University Press, 1981).

Festinger, Leon, Henry W. Riecken, and Stanley Schachter. *When Prophecy Fails: A Social and Psychological Study of a Modern Group That Predicted the Destruction of the World* (London: Harper-Torchbooks, 1956).

Hawking, Stephen. *The Grand Design* (New York: Bantam, 2010).

Hazen, Robert M., and James Trefil. *Science Matters* (New York: Doubleday, 1991).

Heiserman, David. *Radio Astronomy for the Amateur* (Blue Ridge Summit, PA: TAB Books, 1975).

Heuer, Kenneth. *The End of the World* (New York: Rinehart & Company, 1953).

Hey, J. S. *The Evolution of Radio Astronomy* (New York: Science History Publications, 1973).

Hodge, Paul W. *Atlas of the Andromeda Galaxy* (Seattle: University of Washington Press, 1981).

Hodge, Paul W. *Galaxies* (Cambridge, MA: Harvard University Press, 1986).

Islam, Jamal N. *The Ultimate Fate of the Universe* (Cambridge, England: Cambridge University Press, 1983).

Kaufmann, Williams J., III. *Galaxies and Quasars* (San Francisco: W. H. Freeman and Company, 1979).

Kerr, Richard A. "Experts Agree Global Warming Is Melting the World Rapidly." *Science* 338 (2012): 1138.

Kerr, Richard A. "Homegrown Organic Matter Found on Mars, But No Life." *Science* 336 (2012): 970.

Kuhn, Ludwig. *The Milky Way* (New York: John Wiley & Sons, 1982).

Longair, Malcolm S. *The Origins of the Universe* (Cambridge, England: Cambridge University Press, 1991).

Macvey, John W. *Where Will We Go When the Sun Dies?* (New York: Stein and Day, 1983).

Morris, Richard. *The Fate of the Universe* (New York: PEI Books, 1982).

Motz, Lloyd. *The Universe* (New York: Charles Scribner's Sons, 1975).

Parker, Barry. *Colliding Galaxies* (New York: Plenum Press, 1990).

Parker, Barry. *Invisible Matter and the Fate of the Universe* (New York: Plenum Press, 1989).

Piddington, J. H. *Radio Astronomy* (New York: Harper & Brothers, 1961).

Ronan, Colin A. *The Natural History of the Universe* (New York: Macmillan, 1991).

Silk, Joseph. *The Big Bang* (San Francisco: W.H. Freeman and Co., 1980).

Spilka, Bernard, Ralph W. Hood Jr., Bruce Hunsberger, and Richard Gorsuch. *The Psychology of Religion*, 3rd Ed. (New York: Guilford Press, 2003).

Sullivan, W. T., III, ed. *The Early Years of Radio Astronomy* (New York: Cambridge University Press, 1984).

Sutton, Christine, ed. *Building the Universe* (New York: Basil Blackwell Ltd., 1985).

Verschuur, Gerrit L. *The Invisible Universe* (London: The English University Press Ltd., 1974).

Verschuur, Gerrit L. *The Universe Revealed* (New York: Springer-Verlag, 1987).

Weinberg, Steven. *The First Three Minutes* (New York: Basic Books, 1977).

PART FIVE

Evolution and the Principle of Natural Selection

"Alfred Russel Wallace Goes Online." *Science* 338 (2012): 24.

"Ancient Weaponry", *Science* 338 (2012): 861.

Angela, Piero, and Alberto Angela. *The Extraordinary Story of Human Origins* (New York: Prometheus Books, 1993).

"*Australopithecus sediba*: Skeletons Present an Exquisite Paleo Puzzle." *Science* 333 (2011): 137.

Bennett, M. R., J. W. Harris, B. G. Richmond, et al. "Early Hominin Foot Morphology Based on 1.5-Million-Year-Old Footprints from Illeret, Kenya." *Science* 323 (2009): 1197.

Berger, L. R., D. J. Ruiter, S. E. Churchill, et al. "*Australopithecus sediba*: A New Species of Homo-like Australopith from South Africa." *Science* 328 (2010): 198.

Bronowski, Jacob. *The Ascent of Man* (Boston: Little, Brown and Company, 1973).

Burkhardt, Frederick, and Sydney Smith, ed. *The Correspondence of Charles Darwin*, 9 Vols. (New York: Cambridge University Press, 1985).

Campbell, Bernard G. *Humankind Emerging* (Boston: Little, Brown and Company, 1985).

Cox, Allan, ed. *Plate Tectonics and Geomagnetic Reversals* (San Francisco: W. H. Freeman and Co., 1973).

Coyne, Jerry A. *Why Evolution Is True* (New York: Viking, 2009).

Darwin, Charles. *The Descent of Man* (Princeton, NJ: Princeton University Press, 1981).

Darwin, Charles. *The Origin of Species* (New York: Macmillan Publishing Co., 1962).

Dennett, Daniel C. *Darwin's Dangerous Idea* (New York: Simon & Schuster, 1995).

Dockens, Richard. *River out of Eden* (New York: Basic Books, 1995).

Durant, John, ed. *Darwinism and Divinity* (New York: Basil Blackwell Ltd., 1985).

Eiseley, Loren. *Darwin's Century* (New York: Doubleday & Company, 1958).

Eldredge, Niles. *The Monkey Business* (New York: Pocket Books, 1982).

Eldredge, Niles, and Ian Tattersall. *The Myths of Human Evolution* (New York: Columbia University Press, 1982).

Erickson, John. *Plate Tectonics* (New York: Facts on File, 1992).

Fagan, Brian M. *The Journey From Eden* (London, Thames and Hudson Ltd., 1990).

Gibbons, A. "Breakthrough of the Year: *Ardipithecus ramidus.*" *Science* 326 (2009): 1598.

Gibbons, A. "Close Encounters of the Prehistoric Kind," *Science* 328 (2010): 680.

Grant, Peter R., and B. Rosemary Grant. *How and Why Species Multiply: The Radiation of Darwin's Finches* (Princeton, NJ: Princeton University Press, 2008).

Gould, Stephen Jay. *Wonderful Life* (New York: W. W. Norton & Company, 1989).

Humphrey, Nicholas. *A History of the Mind* (New York: Simon & Schuster, 1992).

Johanson, Donald, and James Shreeve. *Lucy's Child* (New York: William Morrow and Company, 1989).

Kearey, Philip, and Frederick J. Vine. *Global Tectonics* (London: Blackwell Scientific Publications, 1990).

Larson, S. "Did Australopiths Climb Trees?" *Science* 338 (2012): 478.

Leakey, Richard. *The Origin of Humankind* (New York: Basic Books, 1994).

McCarthy, Dennis. *Here Be Dragons: How the Study of Animal and Plant Distributions Revolutionized Our Views of Life and Earth* (New York: Oxford University Press, 2009).

McCrone, John. *The Ape That Spoke* (New York: William Morrow and Company, 1991).

Morris, Desmond. *The Illustrated Naked Ape* (New York: Crown Publishers, 1967).

Poirier, Frank E. *Understanding Human Evolution* (Englewood Cliffs, NJ: Prentice Hall, 1990).

Reeder, John. *Man on Earth* (Austin: University of Texas Press, 1988).

Sagan, Carl. *Broca's Brain* (New York: Ballantine Books, 1979).

Sagan, Carl. *The Dragons of Eden* (New York: Ballantine Books, 1977).

Sagan, Carl, and Ann Druyan. *Shadows of Forgotten Ancestors* (New York: Random House, 1992).

Weiner, Jonathan. *Planet Earth* (New York: Bantam Books, 1986).

PART SIX
The Cell and Genetics

Leadbetter, Edward R., and J. S. Poindexter. *Bacteria in Nature* (New York: Plenum Press, 1985).
Moody, Paul A. *Genetics of Man* (New York: W. W. Norton & Company, 1967).
Morgan, Thomas H. *Heredity and Sex* (New York: Columbia University Press, 1913).
Peacocke, Arthur. *God and the New Biology* (San Francisco: Harper & Row, 1986).
Rensberger, Boyce. *Biology* (New York: Fawcett Columbine, 1996).
Rosen, Robert. *Life Itself* (New York: Columbia University Press, 1991).
Wright, Sewall. *Evolution and the Genetics of Populations.* Vol. 1 (Chicago: The University of Chicago Press, 1968).

PART SEVEN
The Structure of the DNA Molecule

Alberts, Bruce, Dennis Brahe, Julian Lewis, Martin Raff, Keith Roberts, and James D. Watson. *Molecular Biology of the Cell* (New York: Garland Publishing, 1994).
Crick, Francis. *What Mad Pursuit* (New York: Basic Books, 1988).
Gribbin, John. *In Search of the Double Helix* (New York: McGraw-Hill Book Company, 1985).
"International Human Genome Sequencing Consortium: Initial Sequencing and Analysis of the Human Genome." *Nature* 409 (2001): 860.
Jasny, B. R., and L. M. Zahn. "A Celebration of the Genome: Parts I-IV." *Science* 331 (2011): 546.
Kornberg, Arthur. *DNA Replication* (San Francisco: W. H. Freeman and Company, 1980).
Pennisi, E. "Microbial Survey of the Human Body Reveals Extensive Variation." *Science* 336 (2012): 1369.
Shapiro, Robert. *The Human Blueprint* (New York: St. Martin's Press, 1991).
Venter, J. C. "The Sequence of the Human Genome." *Science* 291 (2001): 1304.
Watson, James D. *The Double Helix* (New York: Penguin Books, 1968).

ADDITIONAL SOURCES

Adler, Mortimer J., ed. *Encyclopaedia Britannica*, 29 Vols. (Chicago: Encyclopaedia Brittanica, Inc., 1996).
Bernal, J. D. *Science in History.* 4 Vols. (Cambridge, MA: The MIT Press, 1969).
Boorstin, Daniel J. *The Discoverers* (New York: Random House, 1983).

Brennan, Richard P. *Dictionary of Scientific Literacy* (New York: John Wiley & Sons, 1992).

Bronowski, Jacob. *A Sense of the Future* (Cambridge, MA: The MIT Press, 1977).

Burke, James. *Connections* (Boston: Little, Brown and Company, 1978).

Burke, James. *The Day the Universe Changed* (Boston: Little, Brown and Company, 1985).

Campbell, Joseph. *The Power of Myth* (New York: Doubleday, 1988).

Capra, Fritjof. *The Turning Point* (New York: Bantam Books, 1982).

Carnegie Library of Pittsburgh. *The Handy Science Answer Book* (Detroit: Visible Ink Press, 1994).

Clarke, Arthur C. *Profiles of the Future* (New York: Holt, Rinehart and Winston, 1984).

Cohen, I. Bernard. *Revolution in Science* (Cambridge, MA: Harvard University Press, 1985).

Eiseley, Loren. *The Man Who Saw Through Time* (New York: Charles Scribner's Sons, 1973).

Feldman, Anthony, and Peter Ford. *Scientists & Inventors* (New York: Facts on File, 1979).

Gardner, Martin, ed. *Great Essays in Science* (Buffalo, NY: Prometheus Books, 1994).

Gillispie, Charles Coulston. *The Edge of Objectivity* (Princeton, NJ: Princeton University Press, 1960).

Goldstein, Thomas. *Dawn of Modern Science* (Boston: Houghton Mifflin Company, 1980).

Golob, Richard, and Eric Brus, eds. *The Almanac of Science and Technology* (New York: Harcourt Brace Jovanovich, Inc., 1990).

Hagihara, Nobutoshi, ed. *Experience the Twentieth Century* (Tokyo: University of Tokyo Press, 1985).

Hart, Michael H. *The 100* (New York: A&W Visual Library, 1978).

Hutchins, Robert Maynard, ed. *Great Books of the Western World*, 60 Vols. (Chicago: Encyclopaedia Britannica, 1986).

Ingpen, Robert, and Philip Wilkinson. *Encyclopedia of Events That Changed the World* (New York: Penguin Books, 1991).

Knight, David. *The Age of Science* (New York: Basil Blackwell Ltd., 1986).

Kohlstedt, Sally Gregory, and Margaret W. Rossiter, eds. *Historical Writing on American Science* (Baltimore: Johns Hopkins University Press, 1985).

Kuhn, Thomas S. *The Structure of Scientific Revolutions* (Chicago: The University of Chicago Press, 1970).

Medawar, Peter. *The Threat and the Glory* (New York: HarperCollins, 1990).

Moyers, Bill. *A World of Ideas* (New York: Doubleday, 1989).

Patterson, Orlando. *Freedom* (New York: HarperCollins Publishers, 1991).

Porter, Roy, ed. *Man Masters Nature* (New York: George Braziller, 1988).

Ronan, Colin A. *Science* (New York: Facts on File, 1982).

Rose, Steven, and Lisa Appignanesi, eds. *Science and Beyond* (New York: Basil Blackwell Ltd., 1986).

Russell, Bertrand, *Wisdom of the West* (New York: Crescent Books, Inc., 1959).

Seldes, George, ed. *The Great Thoughts* (New York: Ballantine Books, 1985).

Serres, Michele, ed. *A History of Scientific Thought* (Cambridge, MA: Blackwell Publishers Ltd., 1995).

Simonton, Dean Keith. *Greatness* (New York: The Guilford Press, 1994).

Thomas, Lewis. *The Fragile Species* (New York: Charles Scribner's Sons, 1992).

Tipler, Frank J. *The Physics of Immortality* (New York: Doubleday, 1994).

Williams, Trevor I. *Science* (New York: Oxford University Press, 1990).

Zeldin, Theodore. *An Intimate History of Humanity* (New York: Harper-Collins Publishers, 1994).

INDEX

ABOUT THE AUTHORS

David Eliot Brody has been writing professionally since 1976 and has been a practicing lawyer in Denver, Colorado, since 1974. He has written three books (including this one) and has been a lecturer and adjunct faculty member at the University of Colorado and other universities, teaching graduate and undergraduate courses in law, sociology, and political science.

Mr. Brody served as president and a board member of the Colorado Authors' League, served on the board of directors of the National Writers Club, and is a member of the National Association of Science Writers.

In addition to his full-time law practice, David Brody has had a passion for science and the history of science since childhood. His independent research in the 1980s and 1990s on the history of science and the role of science in modern society led to the publication of the first edition of this book in 1997. In the first edition, and in this updated edition, the combination of his science writing and research with his brother Arnold's in-depth knowledge and expertise as a world-renowned scientist in his field presents to the reader the history of science, the biographies of the great scientists, and the science, scientific principles, and concepts—all for the non-scientist to understand.

David Brody lives with his wife and nephew in Greenwood Village, Colorado, and has two sons who are attending college.

Arnold R. Brody, PhD, is David's brother. He is professor emeritus in the Department of Pathology at the Tulane University Medical School in New

Orleans. Formerly, he was the head of the Lung Pathology Laboratory at the National Institute of Environmental Health Sciences, and he was the vice chairman of the Pathology Department at Tulane, where he taught in the medical school and carried out basic science research on the fundamental molecular mechanisms through which inhaled particles cause lung diseases like scarring and cancer. He completed his formal academic career as a professor in the Department of Molecular Biomedical Sciences at North Carolina State University.

Dr. Brody's scientific expertise is in lung anatomy, cell biology, and related fields, including ultrastructural cytology. He is regarded as one of the leading experts in the world in the area of environmental lung disease, and his undergraduate and master's degrees in zoology and anatomy provided him with a broad background in basic biological sciences. He has published more than 150 peer-reviewed scientific articles and 50 book chapters on the findings resulting from his scientific research. His laboratory was, for more than thirty years, recognized as a "small science" leader as described in the "Six Levels of Scientific Achievement" (Figure E-1 on page 360).

Dr. Brody's original articles and reviews concern new discoveries in his field. Through this work, together with his lecturing; participation on boards, committees, and scientific conferences; and his strong interest in the history of science, he knows firsthand the extraordinarily complex process that is required for developing a significant scientific discovery. Thus, as reflected in *The Science Class You Wish You Had*, he has special expertise to write about the greatest scientific discoveries in history and the scientists who made them. He continues to lecture worldwide, and he is married with two daughters and five grandchildren.